Battery Passport

Battery Passport

A Comprehensive Guide to Technology, Policy, and Global Implementation

Xiao Lin
Ilaï Bendavid
Bin Wu
Shengrui Qi
Xue Wang

WILEY-VCH

Authors

Dr. Xiao Lin
Botree Recycling Technologies
No. 99 Jinjhu Avenue
Suzhou, 215128
China

Mr. Ilaï Bendavid
Botree Recycling Technologies
No. 99 Jinjihu Avenue
Suzhou, 215128
China

Mr. Bin Wu
Botree Recycling Technologies
No. 99 Jinjihu Avenue
Suzhou, 215128
China

Mr. Shengrui Qi
Botree Recycling Technologies
No. 99 Jinjihu Avenue
Suzhou, 215128
China

Dr. Xue Wang
Botree Recycling Technologies
No. 99 Jinjihu Avenue
Suzhou, 215128
China

Cover Design: Wiley
Cover Image: This image was generated by Botree staff using ChatGPT

Typesetting: Straive, Chennai, India
Printing: CPI Group (UK) Ltd, Croydon CR0 4YY

C9783527354870_010726

Library of Congress Card No.: applied for

British Library Cataloguing-in-Publication Data A catalogue record for this book is available from the British Library.

Bibliographic information published by the Deutsche Nationalbibliothek The Deutsche Nationalbibliothek lists this publication in the Deutsche Nationalbibliografie; detailed bibliographic data are available on the Internet at <http://dnb.d-nb.de>.

Print ISBN: 978-3-527-35487-0
ePDF ISBN: 978-3-527-85101-0
ePub ISBN: 978-3-527-85100-3
oBook ISBN: 978-3-527-85102-7

The manufacturer's authorized representative according to the EU General Product Safety Regulation is WILEY-VCH GmbH, Boschstr. 12, 69469 Weinheim, Germany, e-mail: Product_Safety@wiley.com.

Contents

About the Authors

Xiao Lin, PhD, is the Founder and CEO of Botree Recycling Technologies. He received his doctorate from the Chinese Academy of Sciences and focuses on battery material regeneration technologies and solutions. He is a member of SAC/TC294, Convenor of ISO/TC 333/AHG2, Co-chair of the GBA CCMAG Steering Committee, Member of the GBA Battery Passport Steering Committee, Operation Agent of IEA HEV TCP Task 48 Battery Swapping, and participant of IEA HEV Task 40 CRM4EV.

Ilaï Bendavid is a Cambridge-trained specialist in environmental policy, energy systems, and sustainable mobility, with professional experience across industry, international organizations, and European institutions. He holds an MPhil in Planning, Growth, and Regeneration from the University of Cambridge; a BSc in Business and Management from Queen Mary University of London; and graduate-level training in environmental, energy, and transport economics at AgroParisTech. His work has addressed European power-market resilience, natural-gas crisis-response strategies, and urban mobility behavior, including published work on e-scooter trip purposes and modal substitution carried out with the University of Cambridge, Transport for London, and Tier Mobility. At the European Commission (DG RTD), he contributed to the valorization of Horizon Europe projects and supported cooperation between regional networks, cities, and industrial stakeholders in the fields of future mobility and urban innovation. His current interests center on battery-recycling policy, circular-economy regulation, the financing of circular-economy infrastructure, and the development of digital systems such as the Battery Passport, which he advances through his role in government affairs and business development at Botree Recycling Technologies.

Bin Wu is Director of sustainability consulting at Botree Recycling Technologies, holding a master's degree in material science from KU Leuven. As an expert in Product Life Cycle Assessment (LCA), he has accumulated years of experience in product carbon footprint research at leading international manufacturing enterprises, with solid expertise in carbon footprint modeling and analysis. Currently, he is deeply engaged in research, analysis of carbon footprint, recycled content, regulation compliance across the lithium battery and new energy industry chains, and has participated in the compilation and review of numerous domestic and international carbon footprint standards, including participating in the expert review of the European Union Joint Research Centre (JRC)-led Battery Regulation carbon footprint calculation rules (CFB-EV for electric vehicle batteries, CFB-INB for industrial batteries), contributing to the formulation of automobile life cycle assessment regulations under the United Nations World Forum for Harmonization of Vehicle Regulations (UN/WP.29), the Global Battery Alliance (GBA) battery greenhouse gas (GHG) calculation rulebook, and the International Lithium Association (ILiA) lithium product carbon footprint guidelines. He has undertaken over 50 sustainability-related consulting projects.

Shengrui Qi is a Sustainability Consultant at Botree Recycling Technologies, China, holding a master's degree in environmental engineering. His main research interests include Life Cycle Assessment (LCA) of the lithium-ion battery value chain, EU circular economy regulations, and sustainability compliance. He serves as a reviewer for the carbon footprint standard developed by the International Lithium Association and has participated in drafting China's national standards for power battery carbon footprint and labeling.

Xue Wang, PhD, is COO and Principal Analyst of Botree Recycling Technologies. She obtained her doctorate in material engineering from KU Leuven. Her research is focused on new metallurgical process technology and waste resource treatment technology, and she is co-manager of IEA HEV-TCP Task 54, Recycled Materials for Electric Vehicles.

Preface

Global management of energy, technology, and materials is undergoing a profound transformation. At the center of this shift is the battery, powering electric vehicles, mobile devices, and the broader energy transition. However, behind this critical tool lies a complex, often opaque system of supply chains, material sourcing, performance monitoring, environmental costs, and end-of-life management.

Today's battery landscape is marked by fast-moving innovation. Manufacturers are deploying advanced chemistries like lithium iron phosphate (LFP) and nickel-manganese-cobalt (NMC), automating gigafactories, and embedding digital battery management systems. Recycling technologies from pyrometallurgy to hydrometallurgy are evolving rapidly, but economic and logistical challenges remain. Market demand is rising sharply, particularly for EV batteries, creating intense pressure on raw material supply chains. At the same time, ethical concerns over labor practices in cobalt mining, environmental degradation from lithium extraction, and lack of transparency throughout the battery lifecycle pose serious risks to sustainability goals.

This is where the Battery Passport enters the picture.

Mandated under the European Union's 2023 Batteries Regulation, the Battery Passport is a digital record that provides detailed information about a battery's origin, composition, performance, carbon footprint, and recycling status. It aims to bring transparency, traceability, and accountability across the entire battery value chain, addressing both environmental and ethical blind spots.

By 18 February 2027, industrial batteries over 2 kWh, Electric Vehicle (EV) batteries, and Light Means of Transport (LMT) batteries (like those in e-bikes and scooters) must carry a Battery Passport to be legally sold in the European Union (EU). This record will be accessible through a QR code printed or engraved on the battery casing or packaging, linking stakeholders including regulators, manufacturers, and consumers to structured, up-to-date data.

The Battery Passport is part of a wider effort to reshape industrial ecosystems through digitalization and circularity, rooted in a legislative arc that includes the 2006 Battery Directive, 2019 European Green Deal, 2020 Circular Economy Action Plan, and the 2022 Ecodesign for Sustainable Products Regulation.

Implementing this system is far from simple. There are countless moving parts: standards for data quality and format, rules on carbon footprint calculation, enforcement mechanisms, software interoperability, security concerns, and global trade impacts. From labeling requirements (due August 2026) to recycled content disclosures (due August 2028), this framework is still evolving, and its success depends on how well interconnected actors understand and act on their roles.

Roadmap Ahead

This book is designed to provide a detailed and critical analysis of the emerging Battery Passport ecosystem, a complex system at the intersection of technology, business, and public policy. We will break down the regulatory pressures and economic drivers behind the transformation of the battery industry, the technological infrastructure that will make the new system possible, and the impact all of this will have throughout the battery value chain.

Our aim is to provide a digestible and critical analysis to complement official reports. Taking an external view, this book offers context, interpretation, and strategic insights to help stakeholders navigate the complexity of the Battery Passport system and understand how its various components interact across technology, policy, and sustainability domains. If you're seeking a deep technical dive, we recommend consulting the European Union's ongoing regulatory work, particularly the *DIN DKE SPEC 99100*, which outlines the core requirements for Battery Passports under the new EU Batteries Regulation as well as the detailed reports published by the Battery Pass project. These resources provide the official technical specifications and implementation guidelines that underpin the regulatory framework.

Who This Book Is For

This book is for industry leaders, policymakers, researchers, and sustainability professionals who are shaping the future of battery systems. Whether you're building new battery technologies, drafting regulatory frameworks, designing Environmental, Social and Governance (ESG) strategies, or managing supply chains, this book offers the insight needed to understand, implement, and scale Battery Passport systems. It provides a shared foundation for anyone working at the crossroads of technology, sustainability, and policy.

What the Book Covers

The chapters ahead unpack the Battery Passport from every critical angle: its standards, technological architecture, regulatory context, economic implications, environmental impact, and future trajectories, offering readers a complete roadmap for understanding and deploying this transformative system.

Acknowledgments

The authors would like to express their deepest appreciation to **Wiley** for their trust in this project and for publishing the book.

We are sincerely grateful to **Emma Sweigart** for her assistance in proofreading and editing and to **Tina Wu** for her continuous support and coordination throughout the development of this work.

We further would like to express our sincere gratitude to the **Global Battery Alliance (GBA)** for its pivotal role in shaping the global discourse on battery sustainability. In particular, we extend our thanks to **Inga Petersen** and **Graham Lee** from the GBA, their pioneering efforts, insights, and guidance have been instrumental in informing the development of this book, providing invaluable inspiration for our exploration of battery sustainability themes. Finally, we extend our respect to the many **researchers and scholars** whose pioneering work has laid the foundations for the understanding and evolution of sustainable battery systems. Your contributions continue to inspire and guide this ongoing journey toward a more transparent and circular future.

1

Introduction to the Battery Passport

1.1 Why the Battery Passport Matters

Global battery demand is skyrocketing, driven by the electrification of transport, expansion of renewable energy, and digital device proliferation. But behind this surge lies a critical challenge: batteries are resource-intensive, environmentally risky, and often built through opaque, fragmented supply chains.

The environmental costs of the rapid increase in battery manufacturing are already visible. Raw material extraction threatens ecosystems and human rights. Recycling rates are still low. And without lifecycle visibility, manufacturers, regulators, and consumers cannot assess compliance, safety, or sustainability; they are flying blind. This industry needs a solution that drives real sustainability.

Enter the Battery Passport. This digital tool creates a full traceable record of a battery's lifecycle from raw material sourcing to production, use, and end-of-life. By integrating identification data, material composition, performance metrics, and carbon footprint disclosures, the Battery Passport aims to bring transparency to one of the world's most complex and opaque product ecosystems.

At its core, the Battery Passport is about accountability. It empowers governments to enforce regulations. It helps companies meet ESG expectations. And it enables consumers and other stakeholders to make informed, ethical choices. By embedding lifecycle data into each battery, it lays the foundation for a circular battery economy where reuse, recycling, and sustainability are not afterthoughts but built-in features.

Global momentum is growing. The European Union will mandate Battery Passports for industrial and EV batteries starting in 2027. Initiatives like the GBA, backed by the World Economic Forum, are standardizing frameworks. Countries like China and South Korea are exploring their own models, with traceability and supply chain security as key motivators.

The Battery Passport arrives at a pivotal moment. Climate goals depend on scaling clean energy storage. Trade tensions hinge on access to critical minerals. And industries are under growing pressure to demonstrate real evidence of sustainable practices, not just promises. In this context, the Battery Passport is a tool that aims to be a strategic instrument for building a transparent, secure, and sustainable battery economy, expanding energy security, and empowering European governments to effectively manage their critical raw minerals supply chains (Figure 1.1).

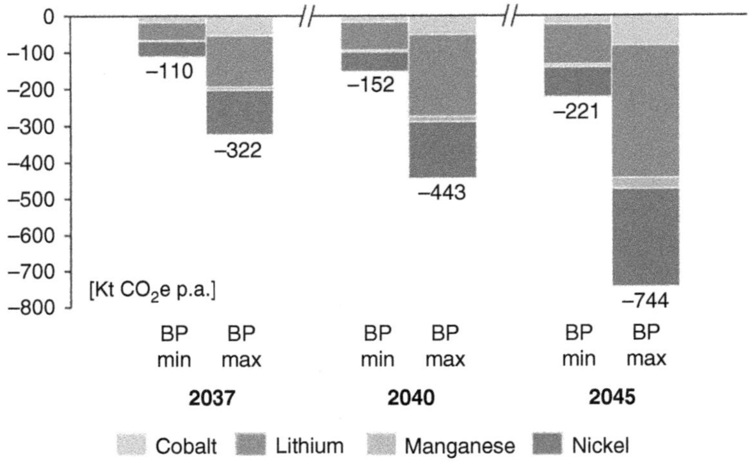

Figure 1.1 CO_2 reduction through primary materials avoided. Due to the additional secondary active materials available from reducing battery leakage, we estimate that **~220–740 kt CO_2 equivalents could be reduced each year** starting in 2045. *Source:* Battery Pass Consortium / Unlocking the Value of the EU Battery Passport. https://thebatterypass.eu/assets/images/value-assessment/pdf/2024_BatteryPassport_Value_Assessment_Synthesis.pdf (accessed 12 November 2025).

1.1.1 Urgency and Strategic Context

Batteries are central to the global transition toward a low-carbon economy, playing a vital role in achieving the ambitious targets of the Paris Agreement and other international climate goals [1]. As global demand accelerates due to electric vehicles (EVs), renewable energy storage, and digital technologies, the battery industry is expanding at breakneck speed.

This is not an abstract or far-off future problem. The speed of deployment is currently outpacing our ability to manage the environmental and social impacts. Without reliable oversight, the battery boom could undermine the very climate goals it is meant to support.

The Battery Passport is designed to help bridge the gap between the demand for batteries and our capacity to produce them responsibly, illuminating blind spots that current systems cannot. This set of digital tools helps users ensure that the materials, processes, and technologies used in battery production and recycling adhere to rigorous environmental standards. By tracking every stage of a battery's lifecycle, battery passports facilitate resource efficiency and align with environmental directives, propelling the global energy landscape toward sustainability and carbon neutrality [2].

Within the EU, the Battery Passport is set to become a mandatory component of the industry by 2027, reflecting the need for a standardized global framework to assess. By facilitating a circular battery value chain, the Battery Passport addresses the significant environmental and social challenges associated with resource-intensive battery production, positioning itself as an essential tool in the global energy transition [1, 3].

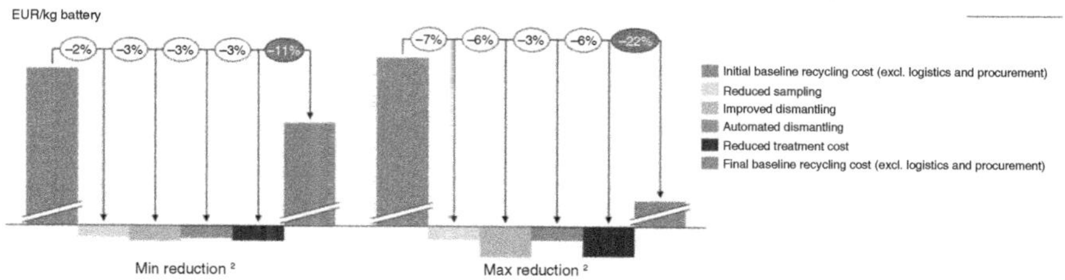

Figure 1.2 Example High-Nickel NMC (622) EV Battery; generic mechanical–hydrometallurgical recycling cost (excluding cost of logistics and procurement). Note that LFP battery recycling has different unit economics. However, the general pre-processing cost reduction levers could apply similarly. *Source:* Battery Pass Consortium / Unlocking the Value of the EU Battery Passport. https://thebatterypass.eu/assets/images/value-assessment/pdf/2024_BatteryPassport_Value_Assessment_Synthesis.pdf/last (accessed 12 November 2025).

Product passports and their consequences are becoming market expectations. They can bring trust in the market, enable smarter regulation, and drive alignment with a future that is low-carbon, circular, and accountable by design (Figure 1.2).

1.1.2 The Problem: Lifecycle Blind Spots and ESG Failures

Despite their growing importance, today's battery production systems operate with serious blind spots. Most supply chains are fragmented and opaque, making it difficult to track the origin of raw materials, monitor production impacts, or ensure ethical end-of-life practices. These gaps expose manufacturers, governments, and investors to significant ESG risks.

Without full lifecycle visibility, companies are more likely to rely on suppliers linked to environmental degradation, forced labor, or illegal mining. Manufacturers are often in the dark about these practices, and there is little incentive to avoid or divest from unethical suppliers due to the plausible deniability created by opaque supply chains. Recycling rates are low, and material recovery is inconsistent, undermining the circular economy goals batteries are meant to support. This lack of accountability poses ethical and environmental challenges, as well as real financial and regulatory consequences. Investors are increasingly divesting from companies that fall short of ESG benchmarks, while regulators are heightening scrutiny through laws that demand traceability and sustainable sourcing all over the world.

The battery industry faces serious challenges with opaque supply chains and limited oversight, especially in the sourcing of critical materials like cobalt and lithium, which are often linked to environmental harm and human rights violations. Stakeholders currently lack the data needed to trace these impacts or hold manufacturers accountable.

Today's battery systems operate in the dark: data is fragmented, supply chains are murky, and accountability is rare. This lack of transparency prevents manufacturers, regulators, and consumers from fully understanding the environmental and ethical costs tied to battery production. As a result, sustainability efforts are often reactive, inconsistent, and disconnected from real-world impacts (Figure 1.3).

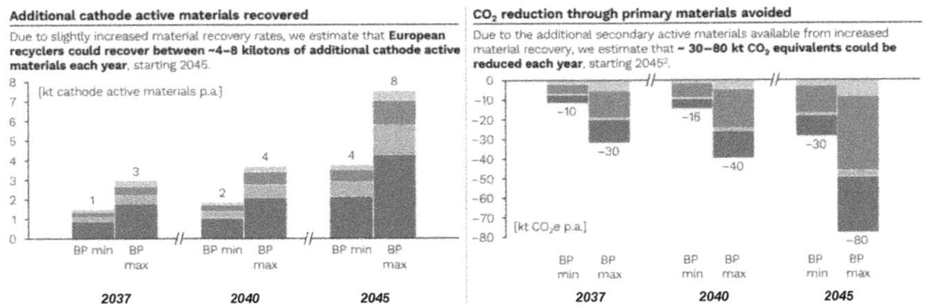

Figure 1.3 Additional cathode active materials recovered. Due to slightly increased material recovery rates, we estimate that **European recyclers could recover between ~4 and 8 kt of additional cathode active materials each year,** starting 2045. **CO_2 reduction through primary materials avoided.** Due to the additional secondary active materials available from increased material recovery, we estimate that **~30–80 kt CO_2 equivalents could be reduced each year,** starting 2025. *Source:* Battery Pass Consortium / Unlocking the Value of the EU Battery Passport. https://thebatterypass.eu/assets/images/value-assessment/pdf/2024_BatteryPassport_Value_Assessment_Synthesis.pdf (accessed 12 November 2025).

1.1.3 The Solution: What the Battery Passport Delivers

As a part of the larger DPP system, the Battery Passport enables compliance with emerging global standards such as the European Union's Battery Regulation and the Ecodesign for Sustainable Products Regulation (ESPR), which aim to ensure traceability, transparency, and circularity in product systems.

By linking each battery to its full lifecycle profile, these passports collect and consolidate the data that matters to stakeholders. Manufacturers gain transparency throughout the supply chain, enabling better risk management and responsible sourcing. Regulators can monitor compliance with evolving environmental and social standards. And consumers receive clear, comparable information that supports informed, ethical purchasing decisions.

This tool transforms the battery from a black box into a traceable, accountable asset. It helps companies future-proof operations against ESG risk, meet market expectations, and actively contribute to decarbonization. In doing so, the Battery Passport accelerates the transition from linear consumption to a circular economy where sustainability is not a trade-off but a competitive advantage (Figure 1.4).

1.1.4 Momentum and Policy Alignment

The Battery Passport is becoming a fast-maturing global standard. Introduced by the GBA at the World Economic Forum, the Battery Passport proof of concept marked a turning point for how the battery industry approaches sustainability, compliance, and lifecycle transparency.

This initiative has garnered widespread support, receiving endorsements from influential global entities, including the G7, the European Union (EU), Canada, and the United States. Within the EU, the Battery Passport is set to become a mandatory

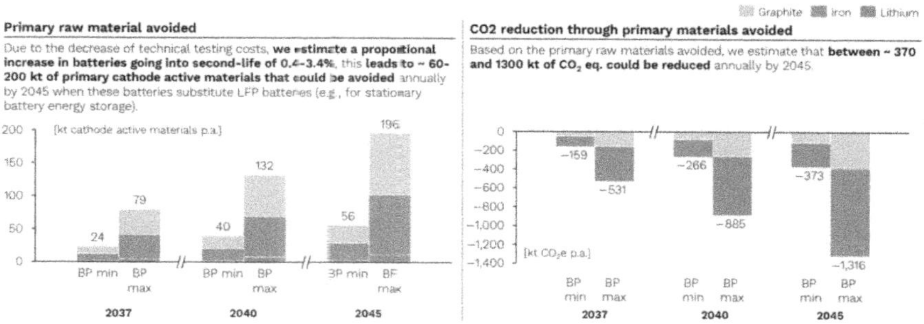

Figure 1.4 Primary raw material avoided. Due to the decrease of technical testing costs, **we estimate a proportional increase in batteries going into second life of 0.4–3.4%**; this **leads to ~60–200 kt of primary cathode active materials that could be avoided** annually by 2045 when these batteries substitute LFP batteries (e.g., for stationary battery energy storage). **CO_2 reduction through primary materials avoided.** Based on the primary raw materials avoided, we estimate that **between ~370 and 1300 kt of CO_2 eq. could be reduced** annually by 2045. *Source:* Battery Pass Consortium / Unlocking the Value of the EU Battery Passport. https://thebatterypass.eu/assets/images/value-assessment/pdf/2024_BatteryPassport_Value_Assessment_Synthesis.pdf (accessed 12 November 2025).

component of the industry in Europe by 2027, reflecting the need for standardized global frameworks to assess and ensure sustainability. By facilitating a circular battery value chain, the Battery Passport addresses the significant environmental and social challenges associated with resource-intensive battery production, positioning itself as an essential tool in the global energy transition [1, 3].

The policy momentum behind the Battery Passport is closely tied to the EU's Battery Regulation Amendment and the broader ESPR. These frameworks are designed to keep hazardous industrial waste out of the environment, improve material recovery, and push producers toward circular economy models. The Battery Passport plays a vital role in documenting extensive lifecycle data, from production to end-of-life management, ensuring regulatory compliance and building consumer trust in battery safety and sustainability standards [4].

To support this transformation, blockchain technology is playing an increasingly central role. It offers a tamper-proof, secure infrastructure for storing and verifying data across complex global supply chains. Supported by initiatives like HORIZON-CL5-2023-D2-02-03 under Horizon Europe, blockchain solutions enhance data interoperability, prevent duplication, and ensure low energy consumption, all while protecting sensitive information. These advancements establish the Battery Passport as a cornerstone of sustainable battery production and management [5].

With the push for transparency and traceability no longer confined to Europe, early adopters of the Battery Passport stand to gain significant advantages, from smoother compliance with emerging regulations to stronger ESG performance and competitive differentiation in sustainability-conscious markets. Countries like China and the United States are actively exploring their own frameworks for battery data governance, supply chain security, and sustainability enforcement. And as with other consumer standards, like the EU's mandate for USB-C charging ports, European regulation often sets the pace globally. Because the EU represents one of the world's largest consumer markets, manufacturers worldwide frequently align with its standards to maintain access. This creates a ripple effect: what starts as an EU requirement often becomes a de facto global norm. This convergence is accelerating, signaling the rise of a new international standard for product accountability.

1.1.5 The Bigger Picture: A Platform for Global Transformation

The Battery Passport is part of a shift in how the world approaches product sustainability, digital governance, and circular economy systems. Its adoption reflects a growing global consensus: to reach climate targets and protect finite resources, we need full lifecycle accountability embedded into every product we make and use.

The EU's ESPR, introduced in 2022, mandates the implementation of DPP across various product sectors. This regulation, which will affect a range of products from textiles to batteries, requires detailed lifecycle data, traceability, and environmental impact assessments, all aimed at ensuring better product management within a circular economy. The Battery Passport fits within this initiative, offering the necessary transparency and detailed data to support the sustainable and circular management of batteries [6].

A new digital infrastructure for sustainable industry, in which every product carries a verifiable record of its environmental and social impact, is emerging. Battery Passports, as an early example of this infrastructure, show how machine-readable data can drive compliance, inform consumers, and streamline global supply chains. They help align disparate stakeholders around shared sustainability metrics and provide a model for other sectors to follow.

Beyond the EU, the principles behind Battery Passports have an influence on global policy design. As major economies explore their own versions of DPPs, interoperability and cross-border standards will become essential. The integration of blockchain and secure data architectures ensures that this transformation is scalable and resilient, building trust into the foundation of trade, technology, and environmental governance.

The Battery Passport is proof that sustainability does not have to come at the expense of complexity but can thrive within it. The right tools can help rewire industrial systems to serve long-term planetary goals.

By establishing a framework for tracking, verifying, and sharing product-level sustainability data, it creates the foundation for a digital infrastructure that can be adapted across nearly every industry. From textiles to electronics to construction materials, the same principles of transparency, traceability, and lifecycle accountability can drive better environmental and social outcomes at scale. What begins with batteries could soon reshape how the entire global economy designs, regulates, and manages products. It is the start of a universal system for embedding accountability into the flow of trade, technology, and materials worldwide (Figure 1.5).

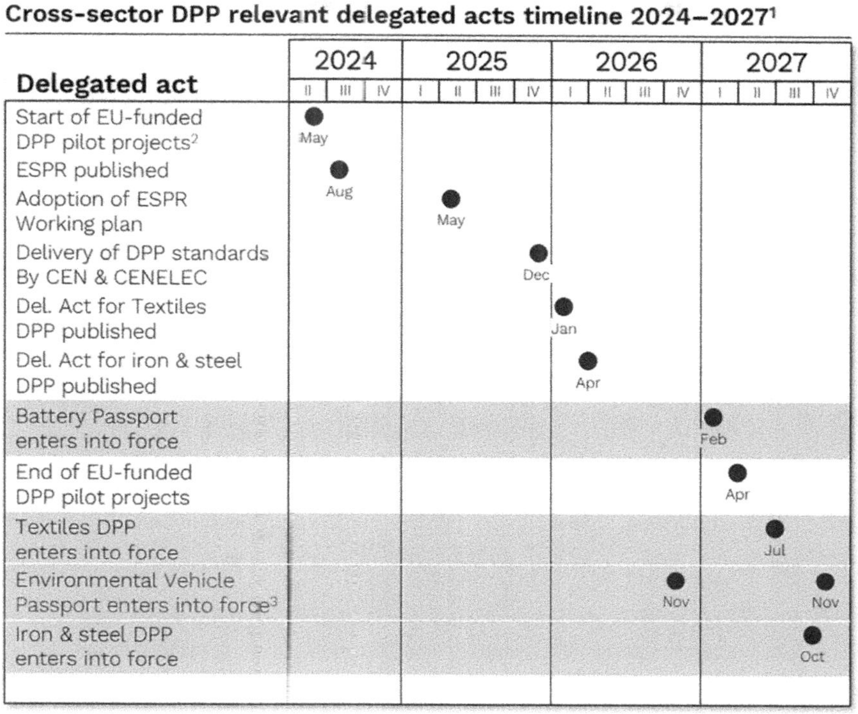

Figure 1.5 Cross-sector DPP relevant delegated acts timeline 2024–2027. *Source:* Adapted from BatteryPass.

1.2 What It Is and How It Works

1.2.1 Core Components

The Battery Passport is a digital record that captures the full lifecycle of a battery from raw material sourcing and manufacturing through use, reuse, and eventual disposal. It provides standardized, verified data that empowers stakeholders across the battery value chain to ensure safety, performance, sustainability, and regulatory compliance.

To make the data meaningful and actionable, the passport is built on seven essential components, grouped into three core categories (Figure 1.6).

1.2.1.1 Product Identity

This group establishes the unique identity and technical foundation of each battery unit.

Identification Data: Each battery is assigned unique identifiers, such as serial numbers and QR codes. These identifiers provide access to digital records and ensure traceability throughout the battery's lifecycle, enabling efficient tracking and management [7].

Manufacturing Information: Detailed records are maintained regarding the battery's production, including the place and date of manufacture, the manufacturer's identity, and physical characteristics such as the battery category

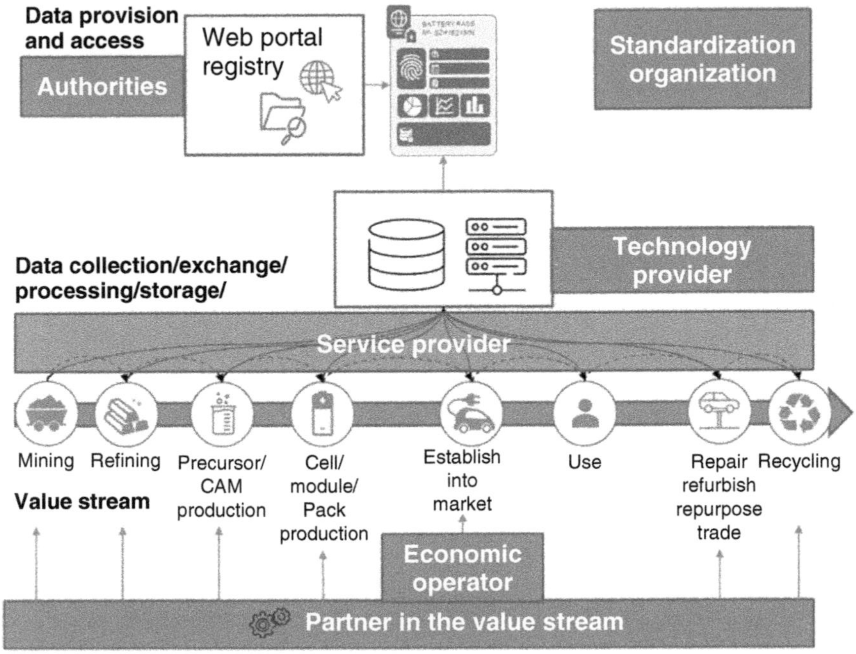

Figure 1.6 The EV battery value stream and the related targeted audience. *Source:* Adapted from Berger et al. [7].

and weight. This data is essential for understanding the production context and ensuring transparency in the supply chain [7].

Technical Specifications: The Battery Passport includes critical technical data, such as the battery type, capacity, energy density, and other performance-related metrics. This information helps determine the suitability of the battery for various applications and assess its performance characteristics [7] (Table 1.1).

1.2.1.2 Lifecycle and Performance

These components track real-world usage and assess environmental and safety standards throughout the battery's operational life.

Lifecycle Data: The passport provides detailed records of the battery's usage, including charging cycles, aging characteristics, and state of health. This information plays a pivotal role in assessing the battery's expected lifetime and efficiency, as well as guiding decisions on repurposing or recycling at the end of its life [4]. However, this should be understood as an objective of the Battery Passport framework rather than a guaranteed outcome, since its practical value will depend on still-evolving rules on access rights, update frequency, standardization of dynamic indicators, and the extent to which commercially sensitive use-phase data can be shared with second-life and recycling actors.

Sustainability and Safety Certifications: Compliance with environmental regulations and safety standards is comprehensively documented in the Battery Passport. Certifications indicate the carbon footprint and recycled content of the battery, offering valuable insights into its environmental impact [4].

Table 1.1 Required battery label data elements, standard sub-clauses, and EU Battery Regulation references.

Information to be displayed on the label	Subclause	BattReg reference
The carbon footprint and the carbon footprint performance class	6.2.4 Carbon footprint label; 6.3 Battery carbon footprint	Article 7(2) via Annex XIII (1c)
Information identifying the manufacturer; • name; • registered trade name or registered trademark; • postal address, indicating a single contact point; • web address, if available; • e-mail address, if available.	6.1.2.4 Manufacturer identifier and information	Article 13(1) via Annex VI Part A (1)
Information identifying the battery: • model identification; • batch or serial number; OR • product number; OR • element allowing identification.	6.1.2.2 Battery identifier	Article 13(1) via Annex VI Part A (2)
Battery category	6.1.3.5 Battery category	Article 13(1) via Annex VI Part A (2)
Place of manufacture geographical location	6.1.3.1 Manufacturing place	Article 13(1) via Annex VI Part A (3)
Date of manufacture (month and year)	6.1.3.2 Manufacturing date	Article 13(1) via Annex VI Part A (4)
Battery mass	6.1.3.6 Battery mass	Article 13(1) via Annex VI, Part A(5)
Capacity	67.2.2 Rated capacity	Article 13(1) via Annex VI, Part A(6)
Chemistry	6.5.2 Battery chemistry	Article 13(1) via Annex VI, Part A(7)
Hazardous substances present in the battery, other than mercury, cadmium, or lead	6.5.5 Hazardous substances	Article 13(1) via Annex VI, Part A(8)
Critical raw materials present in the battery in a concentration of more than 0.1% weight by weight	6.5.3 Critical raw materials	Article 13(1) via Annex VI, Part A(10)
Usable extinguishing agent	6.2.5 Extinguishing agent	Article 13(1) via Annex VI, Part A(9)
Symbol for separate collection	6.2.2 Separate collection symbol	Article 13(4) via Annex VI, Part B, Annex XIII (1q)
Symbol for elements	6.2.3 Symbols for cadmium and lead	Article 13(5)
QR Code	5.2.1 Access to Battery Passport information	Article 13(6) Annex VI, Part C
CE Marking	6.2.7 EU declaration of conformity	Article 20(1)

Source: DIN DKE SPEC 99100:2025-02 [8].

1.2.1.3 End-of-Life and Sustainability

This final group supports responsible disposal, materials recovery, and alignment with circular economy goals.

Material Composition: Details about the materials used in the cathode, anode, and electrolyte are documented in the passport. It also includes information about critical raw materials and any hazardous substances present. This component is particularly vital for recycling and sustainability efforts, as it supports informed decisions about resource recovery and environmental safety [7].

End-of-life Information: Guidelines for recycling or repurposing the battery are a crucial component of the passport. This section includes data on disassembly processes, the recyclability of individual components, and associated environmental considerations, ensuring responsible end-of-life management and promoting sustainability [4].

This architecture turns each battery into a transparent, data-rich asset. By unifying identity, usage, and disposal records in a single digital profile, the Battery Passport lays the foundation for smarter regulation, ethical sourcing, circularity, and climate accountability (Table 1.2).

1.2.2 System Functionality

The Battery Passport is a functional tool that enables visibility, accountability, and continuous improvement across the battery value chain. Its strength lies in how it operationalizes sustainability, turning complex lifecycle data into actionable insights for manufacturers, regulators, and consumers alike.

1.2.2.1 Traceability and Transparency

The Battery Passport provides visibility into the chemical and material composition of each battery. By specifying the general and detailed composition, hazardous

Table 1.2 Summary of core components.

Category	Component	Description
Product identity	Identification Data	Unique identifiers (e.g., serial number, QR code) for traceability.
	Manufacturing Information	Data on origin, manufacturer, production date, weight, and category.
	Technical Specifications	Key metrics like capacity, energy density, and battery type.
Lifecycle and performance	Lifecycle Data	Usage history, charging cycles, aging, and state of health.
	Sustainability and Safety	Certifications on carbon footprint, recycled content, and regulatory compliance.
End-of-life and sustainability	Material Composition	Details on materials, including critical and hazardous substances.
	End-of-Life Information	Recycling guidelines, disassembly data, and environmental considerations.

substances, components, and build, the passport facilitates recycling and reuse. Additionally, it includes data on the battery's carbon footprint, reflecting the environmental impact of production and use, as well as supply chain due diligence, ethical sourcing, and procurement practices. These features align with regulatory mandates requiring operators to publicly disclose their battery due diligence policies, further driving accountability and sustainability [9].

The lifecycle begins with manufacturing, where the Battery Passport offers unparalleled visibility into production processes. By documenting the materials used, carbon footprint, and supply chain practices, the passport enables stakeholders to identify opportunities to improve efficiency and sustainability. Manufacturers can use this data to refine operations and adopt greener production methods, while consumers can make more informed purchasing decisions. This level of visibility encourages demand for responsibly produced batteries, driving systemic improvements across the industry [9] (Figure 1.7).

1.2.2.2 Circularity and Resource Efficiency

Battery Passports play a pivotal role in advancing the circular economy, establishing a closed-loop system that minimizes waste and maximizes resource efficiency at every stage of the battery lifecycle [9]. The passport supports reuse, repurposing, and design-for-recyclability by providing verified, lifecycle-specific data that guides technical and commercial decision-making. It enables manufacturers to design with end-of-life in mind and allows recyclers to extract maximum value from recovered materials.

1.2.2.3 Policy Alignment and Consumer Empowerment

As major initiatives like the European Union's sustainable battery strategy and China's evolving regulatory frameworks gain traction, the Battery Passport promotes

Reporting to the commission: Regulatory requirements

Who reports to whom?	What needs to be reported?	Reporting timeline
Producers and waste management operators → the competent authorities *(Article 75(1)-(3), (7))*	Quantities of: • batteries first placed on the market, excluding batteries that left the Member States before being sold to end users • waste batteries collected and sent for preparation for re-use or for preparation for repurposing • collected waste exported to third countries for treatment, preparation for re-use, preparation for repurposing, or recycling • collected waste sent to treatment and recycling facilities • batteries collected, and the collection rate achieved by the producer or the waste operator	**6 months[1]**
Waste management operators (for treatment and recycling) → the competent authorities *(Article 75(5))*	Quantities of: • waste batteries received for treatment and recycling • waste batteries sent to preparation for re-use, preparation for repurposing, or the recycling process Information on recycling efficiency and on the amounts of materials recovered from waste batteries, as well as the destinations and final outcomes of the output fractions.	**6 months[1]**
Member States → Commission *(Article 76)*	Quantities of: • batteries first placed on the market • waste batteries collected and sent for preparation for re-use or for preparation for repurposing • batteries collected, and the collection rates achieved Values for the recycling efficiencies and material recovery efficiencies achieved	**18 months[1]**

Commission
- Data collected and reviewed by the commission
- A report assessing the organisation of the data drawn up six months after the first reporting of the data and then, every four years

Figure 1.7 Reporting to the commission: regulatory requirements. *Source:* Battery Passport Content Guidance / Battery Pass consortium.

consistent data standards across jurisdictions. These initiatives aim to standardize data collection and exchange, ensuring seamless integration across the entire battery lifecycle. This infrastructure redefines manufacturing and value creation processes to align with circular economy principles. By equipping consumers with actionable insights, the Battery Passport enables them to make environmentally conscious choices, reinforcing sustainable practices as a fundamental component of the battery industry [9].

1.2.2.4 Driving Innovation and Industry Transformation

Beyond compliance, the Battery Passport serves as a catalyst for innovation. As companies adapt to meet its data and sustainability standards, they are incentivized to develop longer-lasting, more recyclable, and higher-performing batteries. This accelerates research and development in battery technologies, supports competitive advantage, and fuels the next generation of energy storage solutions, impacting everything from consumer electronics to EVs and grid-scale storage.

1.2.2.5 Technical Infrastructure and Digital Backbone

The Battery Passport is powered by a layered digital infrastructure designed to ensure data integrity, accessibility, and interoperability across the battery lifecycle. Its architecture combines physical identifiers with secure, tamper-proof systems to manage sensitive information and support circular economy goals at scale. The passport functions as a digital twin of the physical battery, evolving in real-time as the product moves through its lifecycle.

1.2.2.6 Blockchain Backbone and Data Integrity

Blockchain could be used as the digital backbone of the Battery Passport, ensuring that all recorded data on sourcing, production, usage, and disposal is verifiable and cannot be altered retroactively, according to industry players like Minespider. This is especially important in supporting trust across a complex, multiactor value chain. Blockchain also enables decentralized verification of sustainability claims, which is critical as companies face increasing scrutiny over ESG performance.

1.2.2.7 Infrastructure for Interoperability and Expansion

The digital infrastructure supporting the Battery Passport is built for scalability. It integrates with wider systems like the EU's DPP framework and is designed to operate across jurisdictions. By adopting common standards and shared protocols, it enables seamless data exchange between manufacturers, governments, and recyclers. This foundation also opens the door for advanced capabilities, including real-time monitoring, artificial intelligence (AI)-enabled supply chain insights, and automated compliance reporting (Figure 1.8).

1.2.3 Regulatory and Governance Framework

The Battery Passport operates within a tightly defined regulatory ecosystem shaped primarily by the European Union. At the heart of this system is the EU Battery

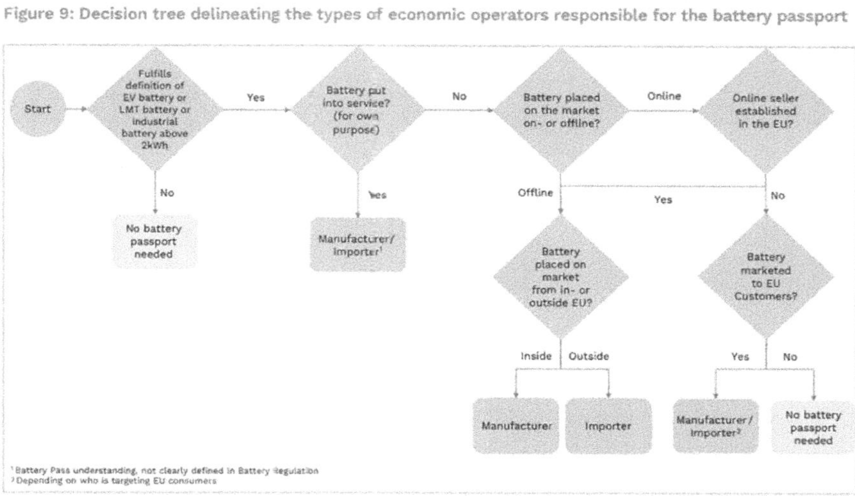

Figure 1.8 Decision tree delineating the types of economic operators responsible for the Battery Passport. *Source:* Battery Passport Content Guidance / Battery Pass consortium.

Regulation Amendment (to ensure a uniform and binding application of the new rules, the previous Directive 2006/66/EC was replaced with an actual regulation: European Regulation 2023/1542), which lays out detailed compliance obligations and performance benchmarks. The Battery Passport is part of this EU Battery Regulation Amendment, being a comprehensive framework designed to mitigate the environmental and social impacts of battery production. By, among others, focusing

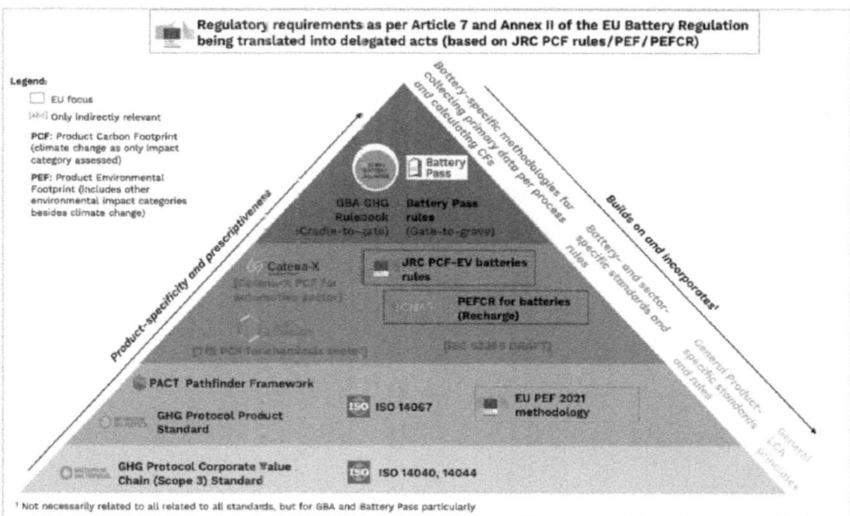

Figure 1.9 Regulatory requirements. *Source:* Batteries European Partnership Association (BEPA) [10].

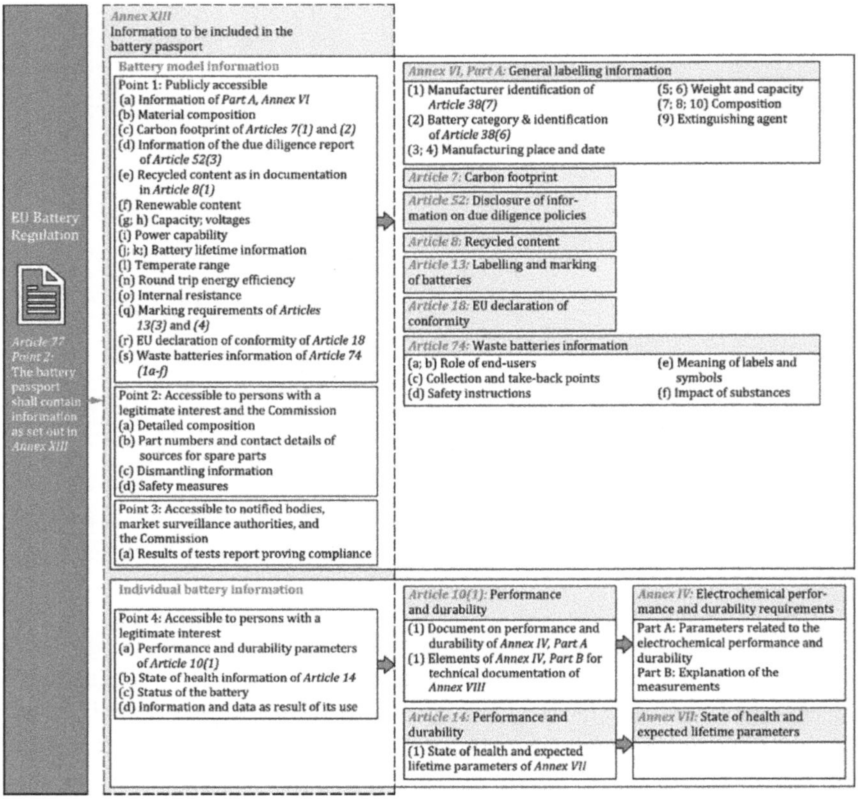

Figure 1.10 Battery Passport information flow in the battery regulation (DIN DKE SPEC 99100 [8]).

on reducing hazardous substances and enhancing recycling rates, the amendment reinforces the principles of a circular economy (Figures 1.9 and 1.10).

As a cornerstone of this initiative, the EU's Battery Regulation specifies rigorous content requirements for the passports. These include comprehensive data on general battery and manufacturer details, compliance certifications, carbon footprints, supply chain diligence, material composition, circularity metrics, resource efficiency, and durability standards. This regulation underscores the EU's commitment to fostering transparency and advancing circular value chains in battery manufacturing [11, 12].

1.3 Comprehensive Overview of the EU Battery Regulation (Regulation 2023/1542)

(Figures 1.11 and 1.12)

Within scope of battery passport

Battery Categories	Battery Definition and Use Cases	Battery Weight
Electric vehicle (EV) battery	• Provide electric power for the traction to hybrid or electric vehicles – of categories L (Regulation (EU) No 168/2013), if larger than 25 kg, or – of categories M, N or O (Regulation (EU) 2018/858)	> 25 kg (category L)
Light means of transport (LMT) battery	• Provide electric power for traction to wheeled vehicles that can be powered by an electric motor alone or by a combination of motor and human power including type-approved vehicles of category L (Regulation (EU) No 168/2013), e.g., e-bikes and e-scooters	≤ 25 kg
Industrial battery[1]	• Designed specifically for industrial uses or • Intended for industrial uses after being subject to preparation for repurposing or repurposing, **or** • Any other battery above 5 kg that is not an LMT, EV nor SLI battery • Industrial uses include, non-exhaustively (*Recital 15*): – industrial activities – communication infrastructure – agricultural activities – energy storage in private or domestic environments – generation and distribution of electric energy – traction in other transport vehicles including rail, waterborne and aviation transport or off-road machinery *Subcategory:* **Stationary battery energy storage system** – industrial battery with internal storage – specifically designed to store and deliver electric energy from and into the grid or store and deliver electric energy to end-user, regardless of where and by whom this battery is being used	> 5 kg (if no other category applies)
SLI (starter, lighting, or ignition) battery	• Designed to supply electric power for starter, lighting, or ignition • May also be used for auxiliary or backup purposes in vehicles, other means of transport or machinery	-
Portable battery	• Not designed specifically for industrial uses • Neither an electric vehicle battery, nor a light means of transport battery, nor an SLI battery	≤ 5 kg
Portable battery of general use	• Rechargeable and non-rechargeable portable batteries specifically produced to be interoperable • Common formats: 4,5 Volts (3R12), button cell, D, C, AA, AAA, AAAA, A23, 9 Volts (PP3) • Providing traction to wheeled vehicles considered as toys (within Toy Safety Directive 2009/48/EC)	-

[1] Only industrial batteries above 2 kWh within scope of battery passport

Figure 1.11 Overview of battery category definitions. *Source:* Battery Pass Consortium [13] / Battery Pass Consortium.

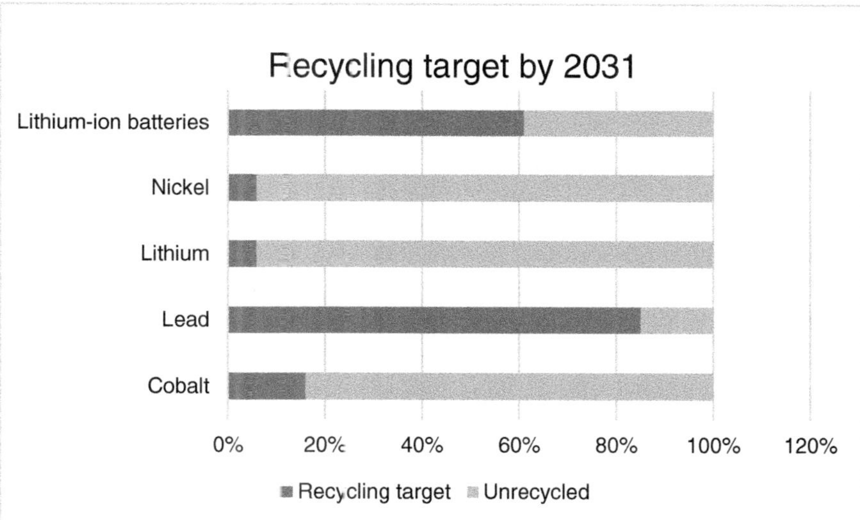

Figure 1.12 Recycling target by 2031.

Recycled Content Targets:

- By 2031, the following percentages of materials must be recovered from battery manufacturing waste or post-consumer waste:
  - Cobalt: 16%
  - Lead: 85%

  - Lithium: 6%
  - Nickel: 6%
- By 2036, these targets increase to:
  - Cobalt: 26%
  - Lead: 85%
  - Lithium: 12%
  - Nickel: 15%

Labeling and Collection Requirements:

- Every battery must clearly indicate the amount of recycled content it contains.
- For Light Means of Transport (LMT) batteries, the collection rate targets are:
  - 45% by 2028
  - 61% by 2031

End-of-life Management:

- There is a total prohibition on the landfilling of waste batteries.
- Economic operators are required to collect end-of-life batteries from end users free of charge.

1.3.1 Manufacturing Liability and Labeling

All EVs, LMT, and rechargeable industrial batteries with a capacity exceeding 2 kWh must include a carbon footprint declaration and label. This label should detail the recycled content of cobalt, lead, lithium, and nickel used in the battery's production. This information must also be accessible via a QR code by 2027.

Additionally, these batteries will require a digital Battery Passport. This passport will contain comprehensive information about the battery model, its intended use, and specific details such as capacity, performance, durability, chemistry, and its recycling collection symbol.

Restrictions on Hazardous Substances

- **Mercury:** The limit is set to no more than 0.0005% by weight.
- **Cadmium:** The limit is set to no more than 0.002% by weight.
- **Lead:** The limit is set to no more than 0.01% by weight in portable batteries by 18 August 2024.

A report will be submitted to the European Parliament and Council by 2027. This report may propose expanding the list of restricted substances to include other materials that are hazardous to human health or that hinder the recycling of safe and high-quality secondary raw materials in batteries.

4:34 p.m.

End Uses and Applications (Figure 1.13)

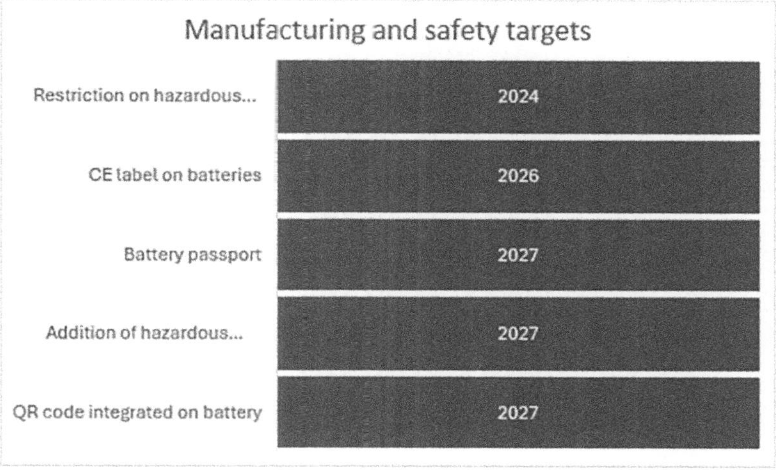

Figure 1.13 Manufacturing and safety targets.

1.3.2 End User Liability

By 2026, batteries must display the "CE" label to show compliance with EU health, safety, and environmental standards. These labels should be prominently placed on the entire device, not just the battery, and must also be accessible via a QR code by 2027.

Users are required to dispose of waste batteries separately from other types of waste.

LMT batteries need to be replaceable by an independent professional, with users having 42 months to adjust their product designs accordingly.

Additionally, Battery Management Systems (BMS) must include a software reset function to facilitate the installation of a different BMS for second-life or end-of-life use. Once this reset is performed, the original manufacturer will no longer be responsible for the battery.

1.3.3 Supply Chain Due Diligence (Updated Timeline)

Due diligence obligations, originally set for 2025, now apply from **18 August 2027**. These apply to any operator placing batteries on the market or putting them into service and cover raw materials such as cobalt, lithium, nickel, and natural graphite.

Requirements Include:

- Public due diligence policy (aligned with Organisation for Economic Co-operation and Development [OECD] and United Nations Guiding Principles on Business and Human Rights [UNGP] standards)
- Risk management systems
- Assessment and mitigation strategies
- Third-party audit by a notified body

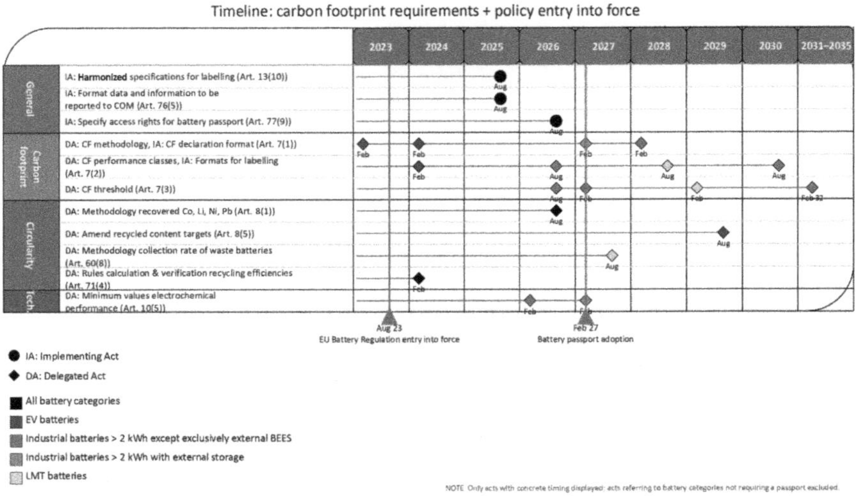

Figure 1.14 Timeline of some battery passport-related delegated and implementing acts following the Battery Regulation. Delegated and implementing acts scheduled by the battery regulation (DIN DKE SPEC 99100 [8]).

Key Changes:

- Threshold raised: Applies only to companies with >€150 million global annual turnover
- Reporting: Every three years starting in 2028
- EU guidance: Delayed to July 2026

Despite the delay, companies must be operationally ready before the 2027 enforcement deadline (Figure 1.14).

The new EU Battery Regulation introduces numerous changes affecting the entire battery lifecycle, including stricter requirements for manufacturing, usage, and recycling. Key aspects include:

- **New Battery Categories: In addition to existing battery types, new categories have been introduced, including batteries for light vehicles (e.g., e-bikes) and traction batteries for EVs.**
- **CO_2 Footprint: From 18 February 2025, manufacturers must disclose the carbon footprint of their batteries. By 2033, CO_2 limits will be introduced.**
- **Increased Collection Targets: The collection rate for portable batteries must rise to 73% by 2030. For light vehicle batteries, targets are 51% by 2028 and 61% by 2031.**
- **Minimum Recycled Content: From 2028, manufacturers must document the recycled material content in batteries. Minimum requirements for cobalt, lithium, nickel, and lead will apply from 2031.**

- **Battery Replaceability: From 2027, consumers must be able to remove and replace portable batteries themselves. Stricter replaceability rules also apply to other battery types.**
- **Extended Labeling Requirements: Batteries must feature QR codes containing essential information on capacity, durability, and recycling. From August 2025, labeling with the "Separate Collection" symbol will be mandatory.**
- **Digital Battery Passport: This will be mandatory for certain battery types from 2027 and will provide information on origin, CO_2 footprint, and material composition.**
- **New Due Diligence Obligations: Companies with revenue exceeding €40 million must implement additional supply chain monitoring measures.**

Battery Waste Management and Producer Responsibility
EPR is strengthened:

- Free battery take-back systems must be available to consumers.
- Distributors must collect used batteries on a **0:1 basis** (no purchase required).

Collection Targets:

- **Portable Batteries:**
  - 45% by 2023
  - 63% by 2027
  - 73% by 2030
- **LMT Batteries:**
  - 51% by 2028
  - 61% by 2031

SLI, industrial, and EV batteries do not have fixed collection targets but still require robust EPR systems.

1.3.4 Mandatory Recycling Efficiency and Recovery Targets

By **31 December 2027**, certified recycling facilities must meet minimum efficiency targets:

- **90%** for cobalt, copper, nickel, and lead
- **50%** for lithium

Recovered materials must be prioritized for reuse in the production of new EV, industrial, and Starting, Lighting, and Ignition (SLI) batteries supporting the EU's circular economy strategy.

1.3.5 Transition Timeline Through 2036

Although key obligations start between 2024 and 2028, the regulation's full scope will continue unfolding until **2036**. This phased implementation allows time for infrastructure development, regulatory clarity, and industry adaptation across all member states (Tables 1.3 and 1.4).

Table 1.3 EU Battery Regulation explained.

Regulated areas	Portable	LMT	SLI	Industrial	EV
CE conformity assessment of batteries based on sustainability[1] and safety[2] requirements	X	X	X	X	X
Supply chain due dilligence for raw materials; cobalt, natural graphite, lithium or nickel[3]	X	X	X	X	X
Battery passport for individual samples	-	X	-	X[4]	X
Removability and replaceability of batteries in products	X	X	-	-	-
Battery waste management	X	X	X	X	X

[1]Different sustainability requirements depending on the battery category
[2]Only applicable for stationary battery energy storage systems
[3]Does not apply to economic operators with a turnover of less than EUR 40 million
[4]Only applicable for industrial batteries > 2 kWh
Source: [14] / *Intertek Group plc.*

Table 1.4 EU Battery Regulation explained.

<table>
<tr><th rowspan="2">Article</th><th rowspan="2">Description</th><th rowspan="2">Portable</th><th rowspan="2">LMT</th><th rowspan="2">SLI</th><th colspan="2">Industrial</th><th rowspan="2">EV</th></tr>
<tr><th>≤2kWh</th><th>>2kWh</th></tr>
<tr><td>6</td><td>Restriction of substances (Cd/Hg/Pb)</td><td>X/X/X</td><td>-/X/-</td><td>-/X/-</td><td colspan="2">-/X/-</td><td>-/X/-</td></tr>
<tr><td>9-10</td><td>Performance and durability requirements</td><td>X[4]</td><td>X</td><td>-</td><td>-</td><td>X[2]</td><td>X</td></tr>
<tr><td>12</td><td>Safety of stationary battery energy storage systems</td><td>-</td><td>-</td><td>-</td><td colspan="2">X[1]</td><td>-</td></tr>
<tr><td>13</td><td>Labelling, marking and information requirements</td><td>X</td><td>X</td><td>X</td><td colspan="2">X</td><td>X</td></tr>
<tr><td>14</td><td>Information on the state of health and expected lifetime of batteries using a BMS</td><td>-</td><td>X[6]</td><td>-</td><td colspan="2">X[1,6]</td><td>X[6]</td></tr>
<tr><td>7</td><td>Carbon footprint</td><td>-</td><td>X</td><td>-</td><td>-</td><td>X[2]</td><td>X</td></tr>
<tr><td>8</td><td>Recycled content (Co/Pb/Li/Ni)</td><td>-</td><td>X[5]</td><td>X[5]</td><td>-</td><td>X[3,5]</td><td>X[5]</td></tr>
</table>

[1]Only applicable for stationary battery energy storage systems
[2]Only applicable for rechargeable industrial batteries > 2 kWh
[3]Only applicable for industrial batteries > 2 kWh with internal storage
[4]Only applicable for portable batteries of general use, excluding button cells
[5]Only applicable if the active material in the electrodes contain Co, Pb, Li or Ni
[6]Only applicable if the battery includes a battery management system (BMS)
Source: Intertek [14].

Select either self-declaration (module A) or involving a notified body (module D1/G)

Mandatory notified body involvement (module D1/G)

1.3.5.1 What Are Producer Responsibility Organisations (PROs) Under EPR in Europe?

***PROs** are approved organizations that act on behalf of multiple battery producers to fulfill their collective obligations under EPR laws. Their role is to implement practical and financial systems for the end-of-life management of batteries.*

Functions of PROs:

1) **Collect and Recycle Batteries**
   - Set up nationwide collection networks.
   - Organize take-back from retailers, public bodies, and private collection points.
   - Ensure environmentally sound recycling (compliant with Art. 70–71 of the EU regulation).
2) **Administrative and Legal Compliance**
   - Report battery volumes to authorities.
   - Monitor and report recycling performance.
   - Handle producers' registration and compliance paperwork.
3) **Financial Management**
   - Collect eco-contributions from producers.
   - Guarantee funding for the collection and treatment of waste batteries.
   - Maintain financial reserves and submit security deposits to protect against insolvency (Batt-EU-AnpG §9).
4) **Ecological Incentivization**
   - Design fee structures to reward recyclability, reusability, and lower environmental impact (Batt-EU-AnpG §10).
5) **Transparency and Public Reporting**
   - Publish annual data on collection rates, fees, ownership structure, and recyclers used (Batt-EU-AnpG §11 Abs. 6).

Who Must Join a PRO?

- **All battery producers** (including importers and online sellers) must either:
  1) Join an authorized PRO, or
  2) Fulfill all obligations individually, which is legally possible but logistically and financially difficult.

1.3.5.2 Role of PROs in the EPR Framework

*They embody the EPR principle by shifting the responsibility and cost of waste management **from taxpayers to producers**, ensuring that the environmental impact of batteries is accounted for across their lifecycle from production to disposal or reuse.*

1.3.5.3 Standards Evolution and Industry Guidance

The CIRPASS initiative, supported by Horizon Europe, contributed to defining key use cases and data requirements for the Battery Passport, helping align its development with real-world stakeholder needs [15].

1.3.6 Integration with DPPs and Global Frameworks

The Battery Passport is a foundational element of the European Union's broader strategy to embed sustainability, traceability, and circularity into product policy through DPPs. As part of the ESPR, DPPs will become mandatory for most product categories sold in the EU starting in 2024. The Battery Passport, with its lifecycle-wide data structure and secure digital integration, exemplifies how DPPs can deliver machine-readable, verifiable insights into product origins, composition, performance, and environmental impact [6].

To provide a comprehensive view of this shift, the EU's Circular Economy Action Plan (CEAP) includes several flagship policies that embed DPPs across industries. These include:

- **ESPR:** making DPPs the central vehicle for transparency on durability, repairability, and resource efficiency.
- **EU Battery Regulation (2026/2027):** requiring Battery Passports for EV and other batteries to ensure supply chain visibility and promote circularity.
- **CSRD and Green Claims Directive:** enabling companies to substantiate ESG claims using verified lifecycle data from DPPs.
- **Sector-specific rules** for textiles, construction, detergents, toys, and electronics, each of which will use DPPs to standardize sustainability reporting and improve product traceability [16].

The success of the Battery Passport is paving the way for this broader rollout. It demonstrates the value of standardized, interoperable digital infrastructure for managing complex product data across international supply chains. The Battery Passport's architecture, including secure blockchain integration, QR code linkage, and real-time updates, serves as a blueprint for other DPP applications.

Lessons learned from its implementation, particularly around data governance, stakeholder coordination, and technical scalability, are now shaping cross-sector standards. By anchoring DPPs in a common framework, the EU aims to streamline compliance, fight greenwashing, and drive innovation through transparent, accountable value chains.

In this way, the Battery Passport is a test case for a unified, data-driven future of sustainable production. As DPPs expand across sectors, the model pioneered by the Battery Passport will help harmonize environmental standards and enable a truly circular, cross-border product economy.

1.3.7 Driving Circularity and Sustainability

(Figure 1.15)

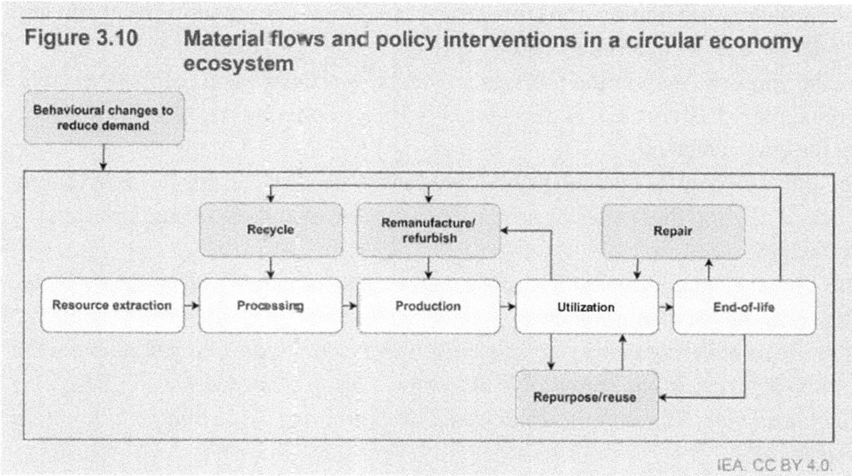

Figure 1.15 Material flows and policy interventions in a circular economy ecosystem. *Source:* Recycling of Critical Minerals / IEA / CC BY 4.0.

1.3.7.1 Enabling Circular Design and Second Life

Short battery lifespan (about seven years), premature disposal, and limited design consideration for reuse have created massive inefficiencies in the battery value chain. Most batteries are treated as disposable once their first-use application ends, even though many retain substantial residual capacity. This not only accelerates resource depletion but also contributes to growing volumes of battery waste that is often difficult or costly to recover due to poor design.

The Battery Passport directly addresses this issue by embedding real-time lifecycle data into each battery's digital identity. This allows stakeholders to monitor usage patterns, performance degradation, and state of health, enabling predictive maintenance and extending the functional life of each unit. Batteries that no longer meet mobility-grade requirements can be repurposed for second-life applications, such as stationary energy storage, where performance demands are lower but longevity and reliability remain critical.

By providing real-time lifecycle data, the Battery Passport enables predictive maintenance, preventing premature disposal and maximizing battery longevity.

The Battery Passport documents the entire lifecycle of each unit, allowing spent batteries to be evaluated for potential reuse or repurposing in second-life applications.

Facilitates repurposing of spent batteries for alternative applications such as stationary storage systems, which extend utility and reduce pressure on raw material supply chains.

However, in many cases, it will likely be more economical and resource-efficient to directly recycle spent batteries rather than repurposing them, particularly as recycling processes become increasingly sophisticated and cost-effective. Recycling can recover valuable materials at higher purity levels, reintegrating them efficiently into new battery production, thus further reducing the environmental footprint and dependency on raw material extraction.

Furthermore, by mandating transparency in battery design and component labeling, the Battery Passport makes disassembly easier, improving recyclability and lowering the cost of recovery.

The Battery Passport encourages manufacturers to adopt design-for-recyclability practices, ensuring that batteries are easier to process at end-of-life and minimizing material loss.

This approach aligns directly with the core principles of the circular economy. As the Ellen MacArthur Foundation notes:

At the heart of the circular economy is the principle that waste and pollution are not inevitable byproducts but preventable outcomes. This model advocates for designing products and materials with the explicit goal of eliminating waste and pollution from the outset [17].

This requires prioritizing durability, reparability, and recyclability in product design, enabling materials to flow through a closed-loop system that mirrors the efficiency of natural cycles. By adopting this proactive approach, the circular economy minimizes environmental impact while maximizing resource efficiency [18].

The circular economy emphasizes the need to keep products, components, and materials in continuous use at their highest utility and value, extending the lifespan of materials, reducing the demand for virgin resources, and minimizing waste [17].

Additionally, the United Nations Development Programme (UNDP) identifies innovative business models, such as product-as-a-service, as key drivers of this transformation [18], which will be explained in more detail in a following chapter (Figure 1.16).

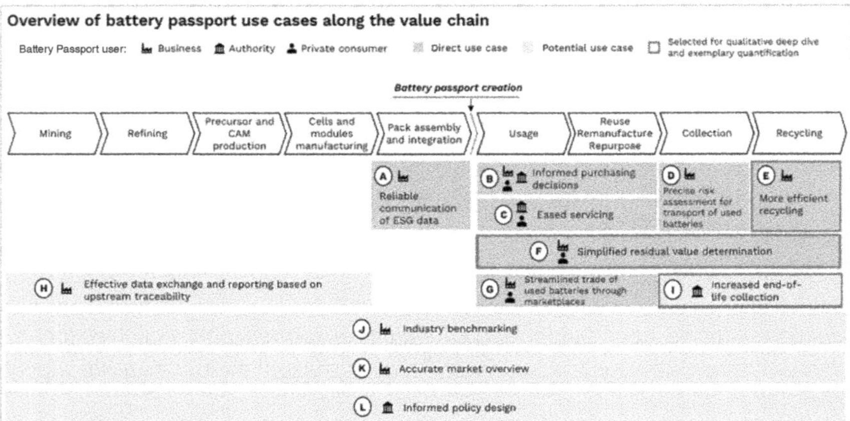

Figure 1.16 Overview of Battery Passport use cases along the value Chain. *Source:* Battery Pass Consortium / Unlocking the Value of the EU Battery Passport. https://thebatterypass.eu/assets/images/value-assessment/pdf/2024_BatteryPassport_Value_Assessment_Synthesis.pdf (accessed 12 November 2025) / CC BY-NC 4.0.

1.3.7.2 Facilitating Recycling and Material Recovery

By acting as a comprehensive digital record that tracks a battery from manufacture to final disposal, it provides the technical detail and historical transparency recyclers need to recover materials efficiently and safely. The passport includes information on design, chemistry, usage history, and performance degradation, giving recyclers a full picture of what is inside before disassembly begins [9].

During a battery's operational life, the passport serves as a repository of real-time performance data, including capacity, energy output, state of charge, and internal resistance. This information allows users to optimize battery performance, extending its usable life and delaying the need for replacements. By maximizing usage, Battery Passports reduce the environmental impact of frequent manufacturing and disposal cycles, aligning with circular economy principles [9].

At end-of-life, this same infrastructure plays a different role: streamlining recovery. With granular, machine-readable data on cathode and electrolyte composition, battery architecture, and hazardous substances, the passport allows treatment facilities to sort and prioritize batteries with precision. This not only improves our ability to increase material recovery yields but also reduces the labor, risk, and uncertainty involved in recycling operations.

When a battery reaches the end of its lifecycle, the detailed information contained in its passport will be a valuable resource for recyclers. The passport provides a comprehensive overview of the battery's composition and history, enabling the efficient recovery of valuable materials such as lithium, cobalt, and nickel. This process minimizes waste sent to landfills and ensures these materials are reintroduced into the production cycle, closing the loop in the circular economy [9].

This closed-loop functionality is increasingly essential as regulatory frameworks evolve. The EU Battery Regulation sets ambitious recovery targets for critical raw materials – targets that will be difficult or impossible to meet without accurate, component-level information. By providing that data up front, the Battery Passport enables recyclers to meet compliance requirements more easily, while helping producers document their contribution to circularity goals.

Ultimately, the Battery Passport serves as a bridge between today's fragmented recovery systems and the high-efficiency, low-waste recycling models demanded by a sustainable energy economy. It turns information into action, powering better decisions, higher yields, and reduced environmental impact at every stage of the battery's end-of-life journey.

1.3.7.3 Advancing ESG and Ethical Supply Chains

Battery manufacturing relies on a complex web of global suppliers, many of which operate in regions with heightened risks of environmental harm, labor violations, and conflict mineral extraction. Without end-to-end visibility, it becomes nearly impossible for manufacturers or regulators to ensure responsible sourcing. This opacity presents growing liability in an era where ESG compliance is no longer optional but expected.

The Battery Passport offers a solution grounded in traceability. By recording supply chain data at each stage of the battery's lifecycle from raw material extraction to component assembly, it enables manufacturers to verify where materials come from, how they are sourced, and whether suppliers meet environmental and social standards.

The Battery Passport drives sustainable material sourcing and ethical practices within the battery manufacturing sector. By enabling traceability of materials from their origin, it ensures transparency across the supply chain, discouraging the use of conflict minerals and promoting environmentally and socially responsible procurement. Manufacturers are held accountable for sourcing practices, advancing industry alignment with global environmental and social sustainability goals [19–21].

This traceability is powered by secure technologies. These tools support due diligence reporting, streamline regulatory compliance, and reduce the risk of greenwashing by corroborating sustainability claims with verifiable data.

Battery Passports also help companies meet mandatory disclosure requirements. For example, supply chain due diligence is increasingly required under EU regulations, including documentation of carbon intensity, hazardous content, and ethical sourcing [9]. With the passport, this data is collected systematically and stored in a format accessible to auditors, investors, and consumers alike.

As ESG reporting becomes a decisive factor in investor confidence and corporate valuation, the Battery Passport gives firms a competitive edge. It identifies ethical gaps and helps close them, enabling a proactive shift toward sustainable, resilient, and socially responsible supply chains.

1.3.7.4 Driving Regulatory and Market Transformation

The Battery Passport has the potential to become a catalyst for structural change in how the battery industry is regulated, funded, and globally aligned. By turning lifecycle data into a standardized and verifiable format, the passport gives policymakers and investors the clarity they need to act decisively.

The passport's granular, traceable data enables the development of new policy instruments. Governments can move beyond static reporting requirements toward dynamic, real-time compliance frameworks. Delegated acts and regulatory amendments like those under the EU Battery Regulation are increasingly being shaped around the capabilities the passport introduces: carbon footprint thresholds, recycled content quotas, and due diligence standards that are trackable, enforceable, and future-proof.

The comprehensive data provided by Battery Passports has a considerable influence on policymaking and investment decisions. The verifiable data from Battery Passports also attract investments in renewable energy infrastructure and innovative battery storage solutions, accelerating the global shift to a low-carbon economy [13].

This policy-aligned data structure also builds investor confidence. With ESG performance now central to financial decision-making, Battery Passports offer a layer of credibility that reduces risk and increases transparency. Lenders, asset managers, and insurers can assess sustainability performance not just from self-reported disclosures but from a tamper-proof digital trail.

The Battery Passport is also accelerating cross-border harmonization. Initiatives like the GBA and the EU-funded Battery Pass project are setting precompetitive

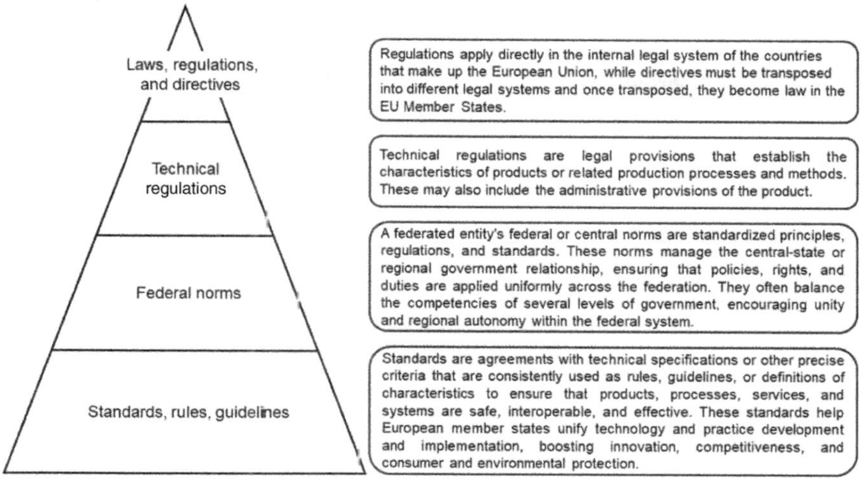

Figure 1.17 Definition of laws, technical regulations, federal norms, and standards. *Sources:* Federal Register [24] and Rufino Júnior et al. [25].

standards that enable interoperability between markets. This is essential for an industry built on global supply chains but increasingly governed by regional policy frameworks. By aligning technical standards with policy requirements, the passport system supports smoother trade, regulatory convergence, and operational consistency across jurisdictions.

The widespread adoption of Battery Passports is likely to influence policy and regulation on both national and international levels. Governments may establish new standards to ensure effective implementation and maintenance of the passport system. These developments could lead to a harmonized global approach to battery management, facilitating international trade while upholding strict environmental and ethical standards. This alignment will ensure that companies remain compliant with evolving regulations while driving policy advancements that prioritize sustainability [21].

Finally, the passport's architecture based on blockchain and interoperable data layers acts as a template for future digital infrastructure. It enables real-time monitoring, automated compliance verification, and AI-driven insights, setting the foundation not just for better batteries but for smarter regulation across the clean tech economy (Figure 1.17).

1.4 How? Setting the Global Standard

1.4.1 EU, US, China, and Global Alignment Challenges

1.4.1.1 The EU: Regulatory Leader and First-Mover

The European Union has emerged as the global frontrunner in regulating battery sustainability, traceability, and circularity. Through a tightly coordinated policy ecosystem, it has positioned the Battery Passport not just as a compliance mechanism but as a driver of industrial transformation.

The Battery Passport aligns closely with the EU Battery Regulation Amendment, a comprehensive framework designed to mitigate the environmental and social impacts of battery production. By focusing on reducing hazardous substances and enhancing recycling rates, the amendment reinforces the principles of a circular economy.

This regulatory push is embedded in the broader CEAP and ESPR, which mandate DPPs across multiple sectors. For the battery industry, this means enforceable obligations backed by a structured, long-term timeline.

The EU Battery Regulation establishes a comprehensive timeline spanning from 2023 to 2040, setting ambitious milestones to advance sustainability and safety across the battery industry. Key phases include (Table 1.5)*:*

The regulation applies tailored requirements to specific battery types:

- *EV batteries*
- *Industrial batteries exceeding 2 kWh (excluding external ESS)*
- *Light mobility (LMT) batteries*

These provisions align with key regulatory articles, such as Article 8(1) for documentation and Articles 8(2) and 8(3) for establishing minimum recycled content requirements [22].

Through this timeline and sector-specific mandates, the EU is setting standards for internal markets, but by doing so, it is influencing global best practices. By embedding digital traceability and lifecycle accountability into law, the EU is establishing the blueprint for how sustainable battery ecosystems should operate worldwide (Figure 1.18).

1.4.1.2 China: Strategic Follower and System Integrator

While the European Union has led the charge in Battery Passport regulation, China has emerged as a fast-moving strategic follower, leveraging its manufacturing

Table 1.5 Battery recycling key phases.

Date	Milestone/event
August 2023	*Regulation enters into force*
February 2027	*Mandatory adoption of Battery Passports for relevant battery types*
August 2027	*Delegated Act adopted for methodologies to calculate recycled content (Co, Li, Ni, Pb)*
December 2027	*European Commission assesses recycled content targets*
August 2030	*Minimum recycled content targets for Co, Pb, Li, and Ni come into effect*
August 2033	*Revised targets + new Delegated Act with documentation rules for recycled content*
August 2036	*Further amended recycled content targets implemented*

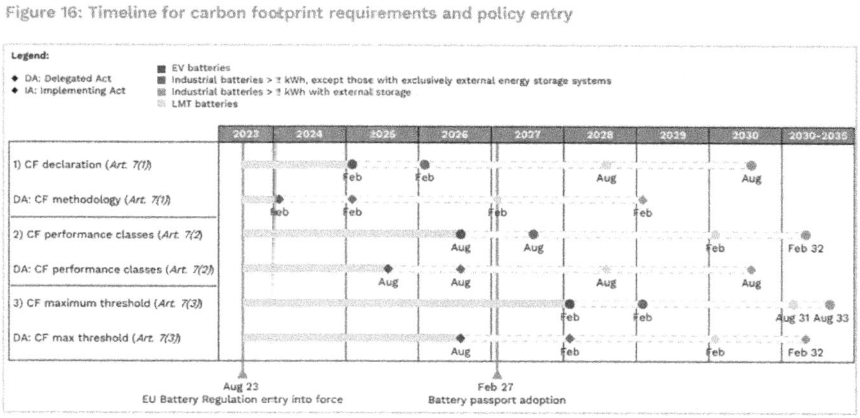

Figure 1.18 Timeline for carbon footprint requirements and policy entry. *Source:* Battery Passport Content Guidance / Battery Pass consortium.

dominance to shape its own framework aligned with global norms. The 2023 update to the EU's Battery Legislation catalyzed similar efforts in China, where policymakers initiated the development of a digital Battery Passport program inspired by the EU's transparency framework. China's initiative prioritizes key aspects such as carbon footprint, circularity, and ESG metrics. By mirroring EU standards, China aims to facilitate smoother trade relations with European partners while ensuring alignment with global sustainability objectives. This growing international momentum highlights the opportunity for policymakers worldwide to standardize reporting parameters, integrate technical frameworks with existing systems, and foster greater engagement from the battery industry [23].

China's regulatory approach is built on a well-established infrastructure. The Interim Provisions on Traceability Management of Power Battery Recycling in New Energy Vehicles, introduced in 2018, mandated end-to-end tracking from cell manufacturers to recyclers. This early legislation laid the foundation for a national digital battery traceability system focused on lifecycle oversight [23, 24].

More recent policies build on that traceability backbone, extending accountability deeper into the supply chain and aligning with international benchmarks. China's evolving Battery Passport framework includes carbon footprint tracking, mandatory lifecycle disclosures, and strict documentation of reuse and recycling activities, all aimed at harmonizing with EU models and enabling cross-border interoperability.

Unlike the more fragmented policy environment in the United States, China has adopted a centralized regulatory stance. National-level rules impose clear obligations on EV battery manufacturers for safe collection, second-life deployment, and end-of-life recycling. These policies go beyond traceability; they also address battery safety, material integrity, and environmental impact throughout reuse cycles [25].

China's strategy includes technical mandates as well. Manufacturers must incorporate reuse- and recycling-friendly design features, including labeling standards and detailed component documentation. This improves transparency and enables more effective disassembly and recovery. In parallel, import/export restrictions have been implemented to prevent the displacement of electronic waste, including used batteries, into lower-regulation regions, reinforcing China's commitment to environmental protection and domestic processing standards [25].

Together, these developments signal China's intent to maintain global competitiveness while enhancing control over its internal battery supply chain. By integrating traceability, sustainability, and trade facilitation into a unified system, China is positioning itself as a systems integrator capable of shaping the next phase of international battery governance.

It is also important to recognize that the overwhelming majority of the world's batteries are either made in China or by Chinese manufacturers. This gives China a unique structural advantage in adopting Battery Passport frameworks: the companies expected to comply with the EU Battery Passport (BP) are, for the most part, already operating under Chinese jurisdiction or influence. As these firms integrate EU-aligned systems into their operations to meet export requirements, the groundwork for broader adoption within China is already being laid regardless of whether domestic policy mandates it. This embedded infrastructure makes it easier for Chinese regulators to scale similar initiatives nationally and aligns market behavior with global norms even in the absence of binding legislation.

1.4.1.3 United States: Fragmented but Evolving

In contrast to the EU's centralized regulatory push and China's integrated strategy, the United States presents a fragmented landscape for battery governance. While the country plays a major role in global EV and battery markets, it lacks a comprehensive federal framework for digital Battery Passports.

In the United States, there is currently no comprehensive federal legislation specifically dedicated to digital Battery Passports comparable to the EU's framework. However, several state-level initiatives and existing federal regulations indirectly contribute to managing end-of-life EV batteries. Federal-level regulation is largely focused on environmental protection, recycling processes, and disposal guidelines. The Resource Conservation and Recovery Act (RCRA) is central in regulating hazardous wastes, including lithium-ion batteries (LIBs). Yet, it lacks explicit guidance or standards tailored specifically to second-life applications or Battery Passports [25].

Instead, regulatory momentum in the United States is driven primarily by a patchwork of state-level EPR initiatives. These frameworks vary significantly by state. California, for example, has adopted stricter rules promoting battery reuse, recycling, and transparency. Still, this decentralized approach creates operational uncertainty for manufacturers and recyclers operating across multiple jurisdictions. Without federal harmonization, Battery Passport implementation remains inconsistent.

Federal efforts have shown potential to shift the narrative. In the United States, recent legislation such as the Inflation Reduction Act emphasizes localizing supply chains and enhancing material traceability. Specifically, the Clean Vehicle Tax Credit, outlined in Section 30D of the Act, mandates detailed provenance data for battery materials to qualify for tax incentives. These efforts demonstrate a growing recognition of the importance of robust traceability and transparency frameworks in ensuring the sustainable production and management of batteries [23, 24]. Though not a full Battery Passport mandate, it represents a policy-level signal that traceability is becoming a prerequisite for economic incentives.

Much of the momentum in the United States has come from the private sector, where major automakers and tech companies are experimenting with their own digital transparency tools, often influenced by EU standards or in preparation for global compliance.

However, the policy outlook remains volatile. The future of federal climate and energy policy is uncertain in the context of the second Trump administration. Key provisions of the Inflation Reduction Act such as traceability-linked tax credits could be delayed, weakened, or repealed. The potential policy reversal raises questions about whether battery passport initiatives will gain traction at the federal level or continue to rely on state action and corporate leadership.

In short, the U.S. Battery Passport landscape is constantly evolving. However, without clear national coordination, its trajectory will likely remain shaped by state policy diversity, market forces, and shifting federal priorities.

Interestingly, even in the absence of a federal mandate, the widespread presence of Chinese batteries in the U.S. market may indirectly accelerate adoption of Battery Passport practices. These companies, already adapting to EU requirements and implementing traceability systems across their global operations, may carry over those practices as standard operating procedure. This habitual compliance could lead to informal alignment with international norms, regardless of domestic policy clarity, embedding passport-like systems into U.S. supply chains through the global habits of the manufacturers involved. This potential ripple effect may have been a factor in the EU's adoption of the Battery Passport.

1.4.1.4 Global Alignment: Challenges and Cooperation

As the Battery Passport moves toward international adoption, global alignment remains one of its most urgent and complex challenges. While the European Union has taken the lead with a binding regulatory framework, other regions, including the United States, China, and various global stakeholders, are still in the early phases of development or alignment. The success of the Battery Passport depends on a unified global effort among industry players, regulators, and other stakeholders. The highly globalized nature of battery value chains necessitates a collaborative approach to designing a scalable, interoperable system. Regional disparities in policies and practices risk undermining the passport's potential unless addressed through coordinated international efforts [26].

1.4.1.5 Challenges: Fragmentation, Interoperability, and Trust

Significant challenges include regional variation in standards, regulatory fragmentation, data silos, and confidentiality concerns. Creating a standardized framework for Battery Passports that is globally accepted poses significant challenges. Regional variations in regulatory standards complicate the development of a universal passport format. Navigating these differences and achieving harmonization is essential to unlock the full potential of Battery Passports [26].

The EU Battery Regulation establishes stringent requirements for the content and functionality of Battery Passports, including detailed manufacturer data, certifications, carbon footprint tracking, and supply chain due diligence. While no significant technological barriers have been identified, subtler issues persist. Ensuring interoperability across various digital systems, safeguarding data privacy and security, and maintaining data accuracy throughout the battery's lifecycle are key challenges. A report in the *Journal of Cleaner Production* highlights the importance of overcoming these barriers through infrastructure investments and the adoption of standardized practices across the battery value chain [27]. Additionally, the need for seamless integration between existing and emerging technologies will require industry-wide collaboration to achieve interoperability and scalability.

Ensuring the accuracy and integrity of data throughout a battery's lifecycle is a core requirement for the Battery Passport's success. Challenges include collecting, storing, and managing data in a tamper-proof manner while maintaining accessibility for stakeholders.

The development and implementation of Battery Passports entail considerable costs, including initial investments in technology for data tracking and storage, ongoing maintenance, and necessary upgrades. Smaller manufacturers may face disproportionate financial burdens, potentially impacting market competition and innovation. Other expenses include training personnel and adapting production lines to meet the new requirements [28].

Widespread stakeholder buy-in is crucial for the successful deployment of Battery Passports. However, manufacturers may resist due to concerns about disclosing proprietary information and the added complexity of compliance. Consumers may worry about data privacy and potential price increases, while recyclers require accurate and actionable data to optimize their operations [29].

Achieving stakeholder acceptance will require clear communication of the benefits, such as improved supply chain transparency, enhanced environmental accountability, and reduced lifecycle costs. Early engagement with all stakeholders and targeted financial and technical support for SMEs will be critical to addressing these concerns and fostering collaboration [29].

1.4.2 Global Standards and Regulations

Battery Passport policies exhibit significant regional and global diversity, shaped by distinct regulatory, market, and technological landscapes. These variations influence how Battery Passports are adopted, impacting trade, market access, and stakeholder collaboration. To address these challenges, international standards bodies like the International Electrotechnical Commission (IEC), the International Organization for Standardization (ISO), and others provide unified frameworks that ensure consistency, interoperability, and sustainability across borders.

1.4.2.1 Global Standards and Regulations

Global standards form the backbone of Battery Passport systems, harmonizing technical, environmental, and trade-related requirements. They enable seamless collaboration between regions while supporting sustainability objectives.

1.4.2.2 IEC Standards: Technical Specifications for Safety and Data

The IEC, among other standardization bodies, establishes technical guidelines for battery systems, focusing on safety, performance, and reliability. These standards, including IEC 62133, contribute to ensuring consistency in safety requirements for secondary cells and batteries, particularly in portable applications. By setting uniform testing and performance criteria, IEC guidelines help reduce safety risks, enhance compliance, and support global interoperability in battery systems.

1.4.2.3 ISO Standards: Environmental Management and Sustainability

ISO standards emphasize environmental responsibility, providing structured frameworks for material traceability, lifecycle emissions monitoring, and circular economy practices. These guidelines align Battery Passport systems with global sustainability goals, such as the UN's Sustainable Development Goals (SDGs), by ensuring responsible resource management.

1.4.2.4 Uniformity and Trade Facilitation

International standards, including those from the IEC and ISO, facilitate global trade by reducing regulatory barriers and standardizing compliance requirements. Harmonized regulations lower administrative costs, streamline cross-border trade, and create a level playing field for companies operating in multiple jurisdictions.

1.4.2.5 Impact on Data Management and Privacy

Global standards also address critical issues of data governance and privacy. For instance, regulations like the General Data Protection Regulation (GDPR) in the EU shape how Battery Passport systems handle personal and business data. These standards provide clear protocols for secure data storage, access, and sharing, ensuring compliance with ethical and legal obligations [30].

1.4.2.6 Supporting Technological Integration

Emerging technologies such as blockchain, the Internet of Things (IoT), and AI highlight the need for global standards. These technologies rely on interoperable systems and secure data formats to function effectively. For example, blockchain can track battery materials securely across borders, while IoT devices monitor performance and lifespan data in real time. IEC and ISO guidelines provide the technical foundation for integrating such advanced technologies, ensuring they operate seamlessly within Battery Passport systems.

1.5 Nonexhaustive List of Standards

To ensure interoperability, traceability, and compliance across the battery value chain, the EU Battery Regulation mandates the development and application of standards governing unique identifiers for Battery Passports. These identifiers must enable consistent identification at various levels of granularity (model, batch, and item) and align with existing Union legislation and global identification frameworks. The standards must support both centralized and decentralized approaches, accommodate diverse existing practices, and ensure semantic and syntactic integrity. Table 1.6 summarizes the most relevant standards and their specific functions.

Table 1.6 Standards for unique identifiers in the Battery Passport system.

Standard	Title/scope	Function in Battery Passport system
ISO/IEC 15459	Automatic identification and data capture – Unique identification	Defines structure for globally unique identifiers across products and supply chains
ISO/IEC 61406	Identification Link	Supports digital links between physical products and online information
ISO/IEC 29161	Unique identification for the Internet of Things	Enables consistent identification of IoT-enabled battery components
ISO/IEC 15418	GS1 and ASC MH10 Data Identifiers	Ensures standardized formatting of identifiers for product categories and packaging
ISO/IEC 9834-8/ RFC 4122	Universally Unique Identifiers (UUIDs)	Defines the generation of globally unique digital identifiers
ISO 17442	Legal Entity Identifier (LEI)	Identifies economic operators legally involved in battery manufacturing and handling
ISO/TR 23249	Blockchain – Identity management overview	Supports identity systems in distributed ledger environments, relevant for decentralized identifiers
ISO/TR 6039	Blockchain – Identifiers of subjects and objects	Structures identifiers for subjects/ objects in blockchain-based tracking
ISO 22383	Authentication solutions for material goods	Provides methods for verifying authenticity of battery data and identifiers
ISO 22385	Trust and Interoperability Framework	Establishes rules for trust mechanisms in cross-system interoperability
ISO 22387	Confirmation procedures for artifact metrics	Offers verification protocols for product metrics and digital seals
ISO 22376	Visible Digital Seal (VDS) data specifications	Defines formats for embedding digital authentication into physical battery labels
ISO 22372	Trustworthy supply chain framework	Guides secure data exchange and identification along the battery supply chain

Table 1.6 (Continued)

Standard	Title/scope	Function in Battery Passport system
ISO/IEC 19762	AIDC – Harmonized vocabulary	Provides a unified terminology for identification and data capture systems
ITU-T X.1403	DLT Security – Secure applications and services	Establishes security requirements for distributed identifier systems
W3C – DID	Decentralized Identifiers (DIDs)	Enables self-sovereign, decentralized identity models for battery items and stakeholders
W3C – Verifiable Credentials	Digital verifiability of credentials	Enables tamper-evident, digitally signed identity claims within the passport system
EN IEC 63365	Digital Nameplate	Standardizes digital product marking, key for linking physical items to digital passports
ISO/IEC 20248	Digital signature schema for AIDC	Ensures secure, machine-verifiable digital signatures embedded in identification codes
ISO/IEC 19845:2015	Universal Business Language (UBL) v2.1	Supports structured business data exchange across supply chain actors
EN 16931 (Parts 1, 3–2, 3–3)	Electronic Invoicing – Core semantics and profiles	Ensures semantic interoperability in invoice data that may be linked to product or passport transactions
ISO 22378	Interoperability of product authentication systems	Facilitates communication across different product ID/authentication systems
ISO 22381	Interoperability of object identification and authentication	Builds frameworks for systems that must operate across independent data environments
CEFACT Cross Industry Invoice	UNECE standard for cross-industry invoice exchange	Supports trusted transactional data related to batteries

To enable traceable, interoperable, and verifiable DPPs, the Battery Passport system must rely on standardized data carriers such as QR codes, Radio Frequency Identification (RFID) chips, or barcodes that link each physical battery to its corresponding digital identity. These data carriers must follow common rules regarding their construction, encoding, durability, and visual identity. They must also contain or reference key types of data: the DPP itself, control data for authentication, and offline-accessible cross-sectoral information. Table 1.7 presents the most relevant existing and proposed standards that serve these functions.

Battery Passports will contain both public and restricted data ranging from technical specifications to sensitive commercial or repair history. To ensure that this information is securely accessed, transferred, and maintained, robust standards are needed for identity management, cybersecurity, and access control. These standards must define how different users (e.g., manufacturers, repairers, recyclers) access specific data elements based on attributes such as role or ownership, while safeguarding

Table 1.7 Standards for data carriers linking physical batteries to digital passports.

Standard	Title/description	Function in Battery Passport system
EN IEC 61406-1	Identification Link	Establishes structure for linking a product to its digital representation
EN IEC 63365	Digital Nameplate – Digital Product Marking	Provides rules for embedding digital identifiers on physical product surfaces
CLC/TR 50489	Feasibility of RFID in EEE for WEEE management	Explores the use of RFID tags for tracking and recycling electronics, applicable to batteries
ISO/IEC 24458	Bar code printer and reader performance testing	Ensures reliable encoding and decoding of barcode data carriers
ISO/IEC 22603-1	Digital representation of product information – General requirements	Lays foundational rules for digital product data formatting
ISO/IEC 21471	Extended rectangular Data Matrix symbology	Specifies a compact, robust barcode format used on small or curved surfaces
ISO/IEC 18004	QR code symbology specification	Defines structure, error correction, and use of QR codes for product data links
ISO/IEC 16022	Data Matrix symbology specification	Standard for 2D matrix codes often used in durable goods like batteries
ISO/IEC 15426-2	Barcode verifier conformance for 2D symbols	Provides conformance criteria for barcode quality
ISO/IEC 15424	Data Carrier Identifiers	Assigns symbology identifiers to distinguish among types of barcodes
ISO/IEC 15415	Print quality test specification for 2D symbols	Establishes metrics for evaluating print clarity and durability
ISO/IEC 15418	GS1 Application Identifiers and ASC MH10 Data Identifiers	Facilitates structured data encoding for supply chain tracking
ISO 23354	Business requirements for visibility of logistics flow	Supports end-to-end tracking, enabling persistent links across logistics phases
ISO 15000-2:2021	AS4 Profile of ebXML Messaging v3	Enables secure business messaging, useful in DPP-linked invoice and logistics data
PEPPOL eDelivery	Secure document transport system for digital procurement	Used for integrating transaction and DPP data
GS1 DataMatrix Guideline	Best practices for implementing GS1-encoded DataMatrix barcodes	Provides encoding structure and implementation guidance
ISO/IEC NP 18975	Encoding and resolving identifiers over HTTP	Defines how digital identifiers can be persistently resolved online
GS1 Digital Link	Embeds GS1 identifiers into URLs	Links physical items directly to online DPPs using globally unique links
W3C DIDs	Decentralized Identifiers	Enables decentralized linking between objects and their digital identities

Table 1.7 (Continued)

Standard	Title/description	Function in Battery Passport system
W3C DID Resolution	DID Resolution Protocol	Specifies how systems can interpret and resolve DIDs into usable digital records
W3C DID Registration	DID registration processes	Defines how new decentralized identifiers can be registered and maintained
Digital Object Identifier (DOI)	Persistent identifier system for digital content	Supports long-term, immutable linking to Battery Passport records
Uniform Resource Names (URN)	Standard for unique and persistent resource naming	Ensures battery DPPs are consistently and uniquely addressable
PEPPOL Service Metadata Locator	Metadata discovery for PEPPOL networks	Helps locate and validate DPP-related data exchange endpoints

integrity, confidentiality, and regulatory compliance. Table 1.8 provides a consolidated view of the existing and proposed standards relevant to these objectives.

Effective implementation of the Battery Passport system depends on seamless interoperability across platforms, actors, and regions. Interoperability must be ensured at three levels: technical (data exchange and system integration), semantic (shared understanding of data meanings), and organizational (harmonized processes and governance models). This requires standardized vocabularies, metadata, data formats, and common information models that allow different stakeholders (manufacturers, recyclers, and regulators) to exchange and interpret data consistently. Table 1.9 presents the core standards guiding this effort.

A robust Battery Passport system relies on seamless, secure, and structured data exchange between various stakeholders (manufacturers, recyclers, regulators, etc.). To enable this, standardization of data exchange protocols, data formats, and data processing rules is essential. These standards define how product passport data is introduced, updated, and transmitted, ensuring system-wide interoperability, auditability, and alignment with the principles of the circular economy. Table 1.10 presents key standards that define these technical foundations.

To maintain long-term access to product passport data, especially for market surveillance and regulatory enforcement, standards must ensure that data remains securely stored, persistent, and accessible, even if the originating economic operator becomes inactive. These standards must support decentralized storage, ensure archival integrity, and allow for trusted retrieval of historical passport versions (Table 1.11).

Table 1.8 Standards for access rights, cybersecurity, and confidentiality in the Battery Passport system.

Standard	Title/description	Function in Battery Passport system
EN IEC 63278-3	Security provisions for Asset Administration Shells	Provides a secure framework for managing asset-specific data, relevant for battery-level passport access
ISO/IEC 29146	A framework for access management	Defines policies and mechanisms to control access to restricted product passport data
ISO/IEC 24760	Identity management framework	Outlines how organizations, devices, and services are uniquely identified and authenticated
ISO/IEC 24761	Authentication context for biometrics	Supports advanced, identity-specific access mechanisms (if biometrics are used in operator environments)
ISO/IEC TS 29003	Identity proofing	Guides on how to securely verify identities before granting passport access
ISO/IEC 20248	Digital signature data structure schema	Ensures data authenticity and tamper-proofing of digital passport entries
ISO/IEC 30147	Trustworthiness methodology for IoT systems	Ensures secure management of IoT-connected battery data and access logs
ISO/IEC AWI 30149	IoT Trustworthiness Framework	Provides standards for evaluating trust levels in systems sharing DPP data
ISO 22385	Framework for Trust and Interoperability	Supports policy-level trust design for system-to-system access coordination
ISO/IEC 27040	Storage security	Ensures secure storage of DPP data in enterprise or cloud environments
ISO 15000-2:2021	AS4 profile of ebXML Messaging v3	Secures DPP-related data exchanges between stakeholders
ISO/TR 6039	Blockchain ID systems	Enables secure object/operator identification on decentralized platforms
ISO 23257	Blockchain reference architecture	Serves as a structural framework for secure DLT-based access control
ISO/TS 23635	Blockchain governance guidelines	Ensures data rights and roles are clearly managed in blockchain environments
ISO/WD TR 23642	Smart contract security best practices	Helps mitigate vulnerabilities when using smart contracts in DPP systems
IEC 62443 series	Industrial cybersecurity	Secures industrial infrastructure handling DPP data
IEC 63069	Security of industrial communication networks	Protects DPP information as it flows through secure communication layers
IEC TR 63283-3 and series	Digital twin/AAS cybersecurity	Addresses security for digital twins linked to product passports

Table 1.8 (Continued)

Standard	Title/description	Function in Battery Passport system
ITU-T Rec X.1144 (XACML v3)	XML Access Control Markup Language	Defines access policies in digital environments, supporting fine-grained passport access control
Verifiable Credentials (W3C)	Digital credentialing for access control	Enables decentralized verification of access rights
ODRL Model & Vocabulary (W3C)	Open Digital Rights Language	Defines machine-readable access policies for data exchange
OAuth2	Secure authorization protocol	Widely used framework for managing user and system access permissions
IETF RFC7515	JSON Web Signature	Defines integrity-protected digital data, used in access tokens and signed DPP payloads
Certificate Transparency (RFCs 6962, 9162)	Public logging of certificate issuance	Ensures visibility and traceability of digital certificate usage in DPP systems
OCSP Stapling (RFC 6066)	Online certificate status validation	Secures and accelerates real-time validation of certificates for authentication

Table 1.9 Standards supporting interoperability in the Battery Passport system.

Standard	Title/description	Function in Battery Passport system
EN IEC 63278	Asset Administration Shell for Industrial Applications	Provides a digital twin model enabling structured data exchange between assets and systems
EN IEC 61360-1	Standard data element types with classification scheme – Definitions and methods	Defines a vocabulary and metadata schema to ensure consistent meaning and data structure
ISO/IEC 21823	Internet of Things (IoT) – Interoperability for IoT systems	Guides integration of IoT devices with Battery Passports through layered interoperability architecture
Baseline Protocol	Protocol for secure and interoperable workflows across companies using blockchain and messaging layers	Ensures business process interoperability using zero-knowledge proofs and private computation
ISO 11354-1:2011	Enterprise Interoperability Framework – Part 1	Provides a framework for aligning business processes, organizational structures, and IT systems

Table 1.10 Standards for data processing, exchange protocols, and data formats in Battery Passports.

Standard	Title/description	Function in Battery Passport system
ISO 9735	EDIFACT syntax rules	Standard for structured electronic data interchange (EDI) across parties
ISO 14533	Long-term digital signature profiles for documents	Enables verifiable and legally compliant data records across time
ISO/IEC 19845	Universal Business Language (UBL)	Standard for electronic business documents (invoices, logistics, etc.)
ISO/IEC 20248	Digital signature data structure for AIDC	Provides machine-verifiable data integrity for barcodes and RFID
EN 16931	Semantic model for electronic invoicing	Ensures standardized and interoperable invoice data linked to DPP records
ISO 23247	Digital twin framework for manufacturing	Aligns passport data with real-time production information
ISO 10303 series	STEP: Product data representation and exchange	Provides comprehensive data formats for technical product information
ISO 15000-2:2021	AS4 Profile of ebXML Messaging	Protocol for secure and standardized B2B message exchange
ISO 23354	Visibility in logistics flow	Ensures supply chain transparency with product-specific data
EN ISO 23386	BIM: Property modeling methodology	Describes and maintains digital properties in standardized data dictionaries
EN ISO 12006-3	BIM: Framework for object-oriented construction data	Organizes construction data for integration with DPPs (especially for battery housing/components)
EN 17549-2	BIM: Configurable construction objects	Defines modular data templates for parts used in battery manufacturing
ISO 59040 (draft)	Circular Economy – Product Circularity Data Sheet	Captures lifecycle attributes supporting repair, reuse, and recycling
ISO/CD TR 6277	DLT – Data flow model	Supports distributed ledger applications for decentralized DPP access and transfer
ISO/AWI TS 23516	DLT – Interoperability framework	Standardizes how distributed ledger systems interconnect for data sharing
PEPPOL eDelivery	Secure e-document transport	Framework for safely exchanging DPP and transactional data between systems

Table 1.10 (Continued)

Standard	Title/description	Function in Battery Passport system
ISO/IEC 19987	EPCIS – Event capture and sharing	Enables standardized event data capture for product traceability
ISO/IEC 19988	Core Business Vocabulary	Defines reusable terms for business events and transactions
IEC 62720	Units of Measurement	Ensures consistency in reporting physical battery characteristics
IEC 61360-4 DB	Common Data Dictionary	Reference database for standardized product descriptors
ECLASS	International standard for classifying and describing products	Supports interoperable product categorization across industries
W3C Web of Things (WoT)	IoT integration framework	Connects physical battery components with DPP systems via semantic web tech
W3C WoT Thing Description (TD)	Device-level metadata specification	Makes battery subcomponents machine-readable and network-interoperable
SAREF (ETSI TS 103 264)	Smart Applications Reference ontology	Provides an ontological model for integrating smart battery and IoT systems
UNECE Cross Industry Invoice	Multisector invoice interoperability	Aligns economic and transaction data with DPP use cases
UNECE SCRDM	Supply Chain Reference Data Model	Cross-industry framework for structuring supply chain data
UNCCL/UNCL/UNLOCODE	UN/CEFACT code lists	Identifiers for logistics, product categories, and geographies
OpenIDConnect/OID4VC/OID4VP	Identity/authentication protocols	Manages secure digital credentialing for DPP actors
W3C Verifiable Credentials	Data model for verifiable digital claims	Enables secure proof of authenticity and ownership for DPP data
JSON-LD/VC-JSON Schema	Semantic web vocabulary formats	Standardizes DPP data representation in linked data contexts
GS1 Digital Link/Data Matrix/GDM	Supply chain linking and traceability standards	Converts product identifiers into web links that resolve to digital passport data
GS1 General Specifications	Barcode and RFID data structures	Governs encoding of key product attributes used in DPPs
Global Traceability Standard	Cross-industry event-level traceability	

Table 1.11 Standards for data storage, archiving, and persistence.

Standard	Title/description	Function in Battery Passport system
EN IEC 63278	Asset Administration Shell for Industrial Applications	Includes data structuring and persistence mechanisms for digital twins and passports
Decentralized Web Node (DWN)	Spec for decentralized, permissioned data storage	Enables distributed, resilient passport storage systems
Encrypted Data Vaults (EDV)	Encrypted cloud storage containers for personal/ enterprise data	Provides secure long-term storage of sensitive DPP information
Certificate Transparency (RFC 6962/9162)	Logs for digital certificate issuance and revocation	Verifies authenticity and traceability of archived passport data and credentials

Table 1.12 Standards for data authentication, reliability, and integrity.

Standard	Title/description	Function in Battery Passport system
ISO/IEC 20248	Digital signature schema for AIDC	Provides trusted, machine-verifiable data signatures for QR codes and identifiers
ISO 22378	Interoperability of product identification systems	Supports authentication across multiple product ID frameworks
ISO 22383	Evaluation of authentication solutions for material goods	Defines testing criteria for anti-counterfeiting technologies
ISO 22385	Framework for trust and interoperability	Supports consistent handling of digitally signed passport data across systems
ISO 22387	Artifact metrics confirmation procedures	Provides verification techniques for embedded physical or digital authentication markers
ISO 8000	Data quality principles	Ensures completeness, accuracy, and consistency of passport datasets
Certificate Transparency (RFCs 6962/9162)	Secure verification of public keys and certificates	Supports nonrepudiation and trust in authentication processes

Battery Passports must be authentic, reliable, and tamper-proof throughout their lifecycle. This requires secure mechanisms to verify data origin, ensure content has not been altered, and authenticate the actors accessing or updating information. Standards here cover digital signatures, verifiable credentials, anti-counterfeiting mechanisms, and logging (Table 1.12).

Standardized Application Programming Interfaces (APIs) are essential for enabling automated passport creation, access, updates, and queries, especially

Table 1.13 Standards for APIs in Battery Passport lifecycle and searchability.

Standard	Title/description	Function in Battery Passport system
REST/HTTP/ SOAP/WS-	Standard protocols for web-based APIs	Defines communication and data exchange patterns (e.g., CRUD operations)
OpenAPI Specification	API documentation and schema standard	Facilitates uniform API development and integration
ISO/IEC 21778	JSON data syntax standard	Defines the structure of JSON payloads for DPP API communications
IETF RFC7515/ RFC7519	JSON Web Signature/Token	Ensures secure authentication and authorization of API users
JSON-LD	Linked data format	Supports semantic richness in API responses
eDelivery/ PEPPOL eDelivery	Secure communication frameworks	Enables trusted API connections between government and economic actors
ISO/IEC 19845	Universal Business Language	Aligns API-based DPP data with transaction and logistics standards
ISO/IEC NP 18975	Encoding/resolving identifiers over HTTP	Ensures links in APIs are persistent and interoperable
OpenID Connect/ OpenID for VC	Authentication and authorization protocols	Provides secure access and identity verification for API users
ISO EN 301549	Accessibility for ICT systems	Ensures APIs meet accessibility standards for public services
JRC APIs4DGov	EU guidance on public sector API design	Provides methodology and best practices for building APIs in digital government systems

from regulators or other third parties. These APIs must be secure, interoperable, and scalable, supporting a range of operations including lifecycle tracking, system integration, and search functionalities across national platforms and registries (Table 1.13).

1.5.1 DIN DKE SPEC 99100: A Key Framework

Incorporating the DIN DKE SPEC 99100 into the global standards discussion adds a critical dimension to the regulatory framework for Battery Passports. This specification, closely aligned with the EU Battery Regulation (EU) 2023/1542, defines the essential data attributes required for digital Battery Passports. These attributes include general battery information, material composition, performance metrics, working conditions in raw material extraction, and carbon footprint. By clearly outlining these elements, DIN DKE SPEC 99100 ensures transparency and sustainability in the battery value chain, helping stakeholders adhere to consistent data reporting and management practices [9].

The integration of DIN DKE SPEC 99100 with existing international standards, such as those from the IEC and ISO, enhances interoperability and data consistency across borders. This harmonized approach facilitates the global adoption of Battery Passports, enabling efficient compliance with diverse regulatory requirements and supporting sustainable practices worldwide. By establishing clear data attribute guidelines, the specification aids companies in implementing the Battery Passport and lays the groundwork for future international standardization efforts. This proactive approach positions the battery industry to adapt to evolving regulatory landscapes and technological advancements, fostering a more sustainable and transparent global battery market [9].

1.5.1.1 Progress: Cooperation, Pilots, and Technical Enablers

Despite these challenges, the global policy and technical landscape are beginning to converge. The GBA advocates for a multistakeholder approach, bringing together businesses, regulators, IT solution providers, and public entities to design a global Battery Passport infrastructure. By enabling secure and efficient data exchange across the battery lifecycle, the GBA aims to foster transparency and standardization, paving the way for a globally recognized system [26].

Projects like Battery Pass, supported by the German Federal Ministry for Economic Affairs and Climate Action, are laying the groundwork for harmonized implementation. The Battery Pass Project is actively working to set standardized requirements and implement them in a demonstrator to align with the EU Battery Regulation [29]. It emphasizes the importance of creating a digital record that documents production conditions, usage history, and essential end-of-life data. This transparency is critical for enabling informed decision-making by end users and recyclers, ensuring that Battery Passport effectively supports sustainable practices [31].

Parallel initiatives like CIRPASS, Catena-X, and programs coordinated through UNEP, UNICEF, and Horizon Europe are also contributing frameworks for DPPs, machine-readable data formats, and blockchain integration. Blockchain technology is especially critical, providing tamper-proof storage and decentralized verification to protect sensitive commercial data while ensuring traceability and compliance across complex, multijurisdictional supply chains.

Additionally, bilateral policy convergence, especially between the EU and China, points to a growing recognition that common standards are not just desirable but necessary. China's recent efforts to build a Battery Passport model inspired by the EU framework underscore the global relevance of harmonized digital systems for carbon reporting, ethical sourcing, and lifecycle oversight.

The global battery economy cannot operate with fractured systems. There is a need for a foundation of interoperability, built on secure digital infrastructure, policy alignment, and industry-wide collaboration. Pilot programs show that alignment is possible. Now, the challenge is scaling it together.

Despite growing momentum and visible pilot efforts, delays in global Battery Passport implementation are highly likely. Aligning diverse regulatory systems, overcoming technical integration hurdles, and securing buy-in across fragmented markets will take time, possibly years longer than current roadmaps suggest. Progress will be uneven, especially where political shifts, infrastructure gaps, or economic pressures slow adoption. Even where rollout proceeds on schedule, not all features of the Battery Passport are expected to be functional from day one. Initial versions may focus on basic traceability and carbon reporting, while more complex capabilities like full lifecycle tracking, interoperability with recycling systems, or integration of ESG metrics could be phased in over time. Stakeholders should plan accordingly, balancing ambition with pragmatism.

References

1 Global Battery Alliance. (2023). Battery passport overview [Internet]. https://www.globalbattery.org/battery-passport/ (accessed 23 March 2025).

2 World Economic Forum (2019). *A Vision for a Sustainable Battery Value Chain in 2030 [Internet]*. Geneva: World Economic Forum. https://www3.weforum.org/docs/WEF_A_Vision_for_a_Sustainable_Battery_Value_Chain_in_2030_Report.pdf.

3 Global Battery Alliance. (2023). Battery passport proof of concept launch [Internet]. https://www.globalbattery.org/press-releases/global-battery-alliance-launches-world%E2%80%99s-first-battery-passport-proof-of-concept/.

4 Circularise. (2023). Battery regulation EU: what you need to know [Internet]. https://www.circularise.com/blogs/battery-regulation-eu-what-you-need-to-know-about-battery-passports.

5 Horizon Europe. (2023). Creating digital passport to track battery materials [Internet]. https://www.horizon-europe.gouv.fr/creating-digital-passport-track-battery-materials-optimize-battery-performance-and-life-validate.

6 Cirpassproject.eu, ESPR Short Summary (38). https://cirpassproject.eu/wp-content/uploads/2023/03/ESPR-short-summary-Final.pdf.

7 Berger K., Baumgartner R. J., Weinzerl M., Bachler, J., Preston K., Schöggl J.P. (2023), Data requirements and availabilities for a digital battery passport: A value chain actor perspective. *Cleaner Production Letters*, 4, 100032, https://doi.org/10.1016/j.clpl.2023.100032. https://www.sciencedirect.com/science/article/pii/S2666791623000052 (accessed 23 March 2025).

8 DIN DKE. DIN DKE SPEC 99100:2025-02 (2025). Requirements for data attributes of the battery passport; Text in English. Berlin: DIN Media, Feb. 2025. 106 p. https://doi.org/10.31030/3582101. https://cdn.prod.website-files.com/68e7c9dbc61a04f046b5fc73/694600f15eea8e026af8dcf9_DIN_DKE_SPEC_99100.pdf (accessed 7 May 2025).

9 Battery Pass Consortium. (2023). Battery passport content guidance. Available from: Battery Passport Guidance.

10 Batteries European Partnership Association (BEPA). Getting the general ID: what projects are supporting the development of the battery passport? [Internet].

Brussels: BEPA. 13 July 2023. https://bepassociation.eu/getting-the-general-id-what-projects-are-supporting-the-development-of-the-battery-passport/ (accessed 13 June 2025).

11 European Commission. Ecodesign for sustainable products regulation. Available from: Ecodesign Regulation.

12 European Union. Regulation (EU) 2023/1542 of the European parliament and of the council. Available from: EU Battery Regulation.

13 The Battery Pass Consortium. (2023). Battery passport content guidance: executive summary. https://thebatterypass.eu/assets/images/content-guidance/pdf/2023_Battery_Passport_Content_Guidance_Executive_Summary.pdf (accessed 28 October 2024).

14 Intertek. EU battery regulation explained [Internet]. 8 November 2024. https://www.intertek.com/blog/2023/08-17-battery-regulation/ intertek.com (accessed 13 June 2025).

15 CIRPASS Project. (2024). Digital product passport use cases report. CIRPASS. https://cirpassproject.eu/wp-content/uploads/2024/04/CIRPASS_D2.2_DPP_UseCases_Report_v2.0.pdf.

16 Protokol. (2024). Digital product passport: complete guide [Internet]. https://www.protokol.com/insights/digital-product-passport-complete-guide/ (accessed 10 January 2025).

17 Ellen MacArthur Foundation. What is a circular economy? https://www.ellenmacarthurfoundation.org/topics/circular-economy-introduction/overview.

18 Climate Promise. What is circular economy and why does it matter? https://climatepromise.undp.org/news-and-stories/what-is-circular-economy-and-how-it-helps-fight-climate-change.

19 Umicore. Battery pass: lighthouse project for battery value chain transparency. https://www.umicore.com/en/newsroom/battery-pass/#:~:text=The%20Battery%20Pass%20aims%20to,and%20proportion%20of%20recycled%20content.

20 Siemens. Battery passport – battery manufacturing. https://www.siemens.com/global/en/industries/battery-manufacturing/battery-passport.html.

21 World Economic Forum (WEF) & United Nations Economic Commission for Europe (UNECE). (2023). Digital battery passport. https://www3.weforum.org/docs/WEF_Digital_Battery_Passport_2023.pdf (accessed 28 October 2024).

22 YRL. (2023). Regulation (EU) 2023/1542 of the European parliament and of the council of 12 July 2023 on batteries and waste batteries. Official Journal of the European Union [Internet]. https://eur-lex.europa.eu/legal-content/EN/TXT/PDF/?uri=CELEX:32023R1542.

23 China Government. Interim provisions on traceability management of power battery recycling in new energy vehicles. Available from: China Battery Recycling (in mandarin).

24 Federal Register. Section 30D new clean vehicle credit. Available from: Clean Vehicle Credit.

25 Rufino Júnior, C.A., Sanseverino, E.R., and Koch, D. (2024). Towards battery digital passport: reviewing regulations and standards for second-life batteries. *Batteries* 10: 115. https://www.mdpi.com/2313-0105/10/4/115.

26 European Centre for Policy Studies. Implementing the EU digital battery passport: opportunities and challenges [Internet]. https://www.ceps.eu/ceps-publications/implementing-the-eu-digital-battery-passport/.

27 Berger, K., Schöggl, J.P., and Baumgartner, R.J. (2022). Digital battery passports to enable circular and sustainable value chains: conceptualization and use cases. *Journal of Cleaner Production* 353: 131492.

28 Battery Pass Consortium. Battery passport value assessment: cost and value perspectives [Internet]. https://thebatterypass.eu/wp-content/uploads/EN_Press-release_Battery-Passport-Value-Assessment.pdf.

29 S&P Global Commodity Insights. battery pass consortium analysis reveals benefits and challenges of European battery passport [Internet]. https://www.spglobal.com/commodityinsights/en/market-insights/latest-news/metals/041124-battery-pass-consortium-analysis-reveals-benefits-challenges-of-european-battery-passport.

30 Global Battery Alliance. GBA battery passport proof of concept. Available from: GBA Proof of Concept.

31 Battery pass consortium battery passport technical guidance. https://www.systemiq.earth/wp-content/uploads/2024/03/2024_BatteryPassport_Technical_Guidance.pdf.

2

Technological Infrastructure of the Battery Passport

2.1 Core Technological Components

The Battery Passport has for objective to be built on a multilayered technological framework that enables the real-time tracking, verification, and management of battery data across its entire lifecycle. Far from being a singular tool, the passport should become an ecosystem of interconnected technologies, each serving a specific purpose in data collection, transmission, analysis, and security. Together, these components form the digital backbone of traceable, transparent, and circular battery value chains.

This section provides a detailed overview of the core technologies powering the Battery Passport system, their technical functions, and the value they offer to stakeholders, including manufacturers, recyclers, regulators, and end users, if it manages to reach all its objectives. Where relevant, we provide deeper technical detail in an embedded appendix-style format to clarify the roles of individual components like sensors, data loggers, and transmission networks (Figure 2.1).

2.1.1 Smart Sensors and IoT Networks

One cornerstone of the envisioned system is the use of smart sensors, embedded in batteries to monitor critical performance metrics such as temperature, voltage, and state of charge. These sensors, connected to IoT networks, would enable real-time data transmission to centralized platforms for processing and analysis. Such capabilities could allow stakeholders to anticipate and mitigate issues such as irregular charge cycles or overheating, thereby extending battery life and optimizing performance [1].

Embedded sensors are the primary data acquisition tools in the Battery Passport architecture. These microelectronic components operate continuously to record a wide range of performance and safety metrics. Their integration with IoT networks allows for low-latency, high-resolution data feeds from individual battery cells to cloud-based analytics platforms.

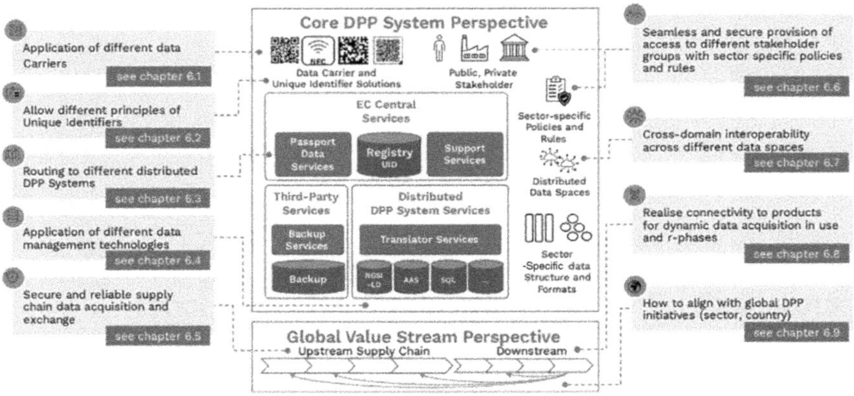

Figure 2.1 Primary interoperability goal: coexistence of principles, standards, and technologies. *Source:* Adapted from The Battery Pass Consortium [1].

Appendix: Types of Sensors and Their Functions

- **Temperature Sensors:**
  Monitor thermal behavior to ensure safe operation. Overheating is one of the leading causes of battery failure and safety incidents, particularly in electric vehicles [2].
- **Voltage and Current Sensors:**
  Measure energy input/output to determine the State of Charge (SoC), assess charging behavior, and flag inefficiencies [2].
- **SoC and State of Health (SoH) Sensors:**
  SoC sensors estimate remaining energy; SoH sensors track degradation trends such as capacity fade and internal resistance over time, offering insights into long-term viability [2].
- **Pressure and Gas Emission Sensors:**
  Detect gas buildup or internal pressure changes that may indicate malfunctions, especially relevant in industrial-scale battery systems [2].

Table 2.1 Useful data that can be read in battery management systems (BMS).

Project	Data
Capacity (Ah or Wh)	For the complete process, the measured capacity, along with total charging and discharging capacity, is measured [3]
Current	The maximum current exceeding the charging and discharging values during the process and the distribution of charging and discharging current during the process, as well as the maximum individual current surpassing the standard results [3]
Voltage (VDC)	For the complete process, the maximum, minimum, and average charging and discharging voltages are required [3]

Table 2.1 (Continued)

Project	Data
Power (W)	Complete power distribution. Full module voltage distribution, with minimum and maximum discharge voltage surpassing the standard values [3]
Remaining power (SOC) (%)	The maximum and minimum remaining power throughout the complete process, as well as the distribution of remaining power at the time of startup and shutdown [3]
Temperature (°C)	The maximum and minimum temperature values as well as the temperature distribution during completion. The maximum and minimum temperature of electronic components and subcomponents in the complete process. The maximum voltage in the complete process
Time (h)	The total time of the production, including the start, usage, and shutdown time. The complete charging, discharging, and balancing time. The complete period of over-voltage, under-voltage, over-high and low temperature, and over-current time during charging and discharging [3]
Internal resistance	Complete measured values [3]
Error message	The variety and quantity of the error messages [3]

Source: Adapted from Rufino Júnior et al. [3].

These data points are vital for ensuring performance, anticipating failure, and enabling second-life assessments (Table 2.1).

2.1.2 RFID and QR Code Integration

To link physical batteries with their digital records, the system could employ RFID tags and/or QR codes. RFID technology, capable of remote identification and tracking, could help manage battery logistics, such as ensuring proper recycling or tracking during transportation. Meanwhile, QR codes would serve as an accessible tool for users, offering instant access to key battery information, such as production details and recycling instructions, through a simple scan. Together, these technologies would make critical information readily available to all stakeholders in the battery lifecycle [1].

- Radio Frequency Identification **(RFID)Tags:**
  Enable noncontact scanning for warehouse, shipping, and recycling logistics. These tags store serial numbers, production dates, material composition, and performance data that can be accessed remotely without requiring line-of-sight.

- **QR Codes:**
  Serve as user-facing access points for the Battery Passport. Printed directly onto the casing of batteries, QR codes can be scanned by smartphones or handheld devices to retrieve full digital records. This ensures transparency at every node from consumer usage to regulatory inspections.

We should note that while QR codes are currently the required standard for linking physical batteries to their digital records, there is ongoing debate within the industry about whether RFID technology might offer a more efficient and scalable alternative *(Article 77(3)), but it leaves the option open for alternative types of smart labels (e.g., RFID, near-field communication (NFC) tags, and Date Matrix Codes) within the scope of delegated acts (Article 13(8)).* RFID could enable automated, contactless data capture throughout the battery lifecycle, which is particularly useful in industrial and recycling contexts. As the Battery Passport and broader Digital Product Passport (DPP) systems mature, manufacturers may be allowed to choose between QR and RFID, depending on their technical and operational needs (Table 2.2 and Figure 2.2).

The choice between QR codes and RFID is shaping up to be a key technical crossroads for Battery Passport implementation. QR codes are simple, low-cost, and universally accessible with a smartphone, making them ideal for consumer-facing transparency and quick manual scans. However, they require direct line-of-sight for both consumers and machines and may degrade over time, especially in harsh operating environments, like cars. RFID, on the other hand, offers durability, remote readability, and automated scanning advantages in high-volume logistics, manufacturing, and recycling operations. But RFID tags are more expensive and

Table 2.2 Comparative overview of RFID tags and QR codes for identification and tracking applications.

Category	RFID tags	QR codes
Technology used	Uses radio frequency technology, utilizing different frequencies for distinct use cases.	Uses optics (lasers); easy to generate on mobile devices or for temporary use.
Read distance	Ranges from a few centimeters to over 100 m, depending on the RFID frequency.	Requires a camera to recognize the code; limited to short distances within direct line of sight.
Tracking capability	Allows tracking and real-time updates via radio signals.	Static: data is transferred only when scanned.
Data capture efficiency	Can capture real-time data from multiple tags simultaneously at different stages.	Can store/transfer large amounts of data via URLs, but must be scanned individually.
Typical solutions/ uses	Example technologies: TRANSIT, uPASS Target, and uPASS Reach.	Example technologies: Nedap's NVITE (reads RFID, QR codes, and more in one reader).

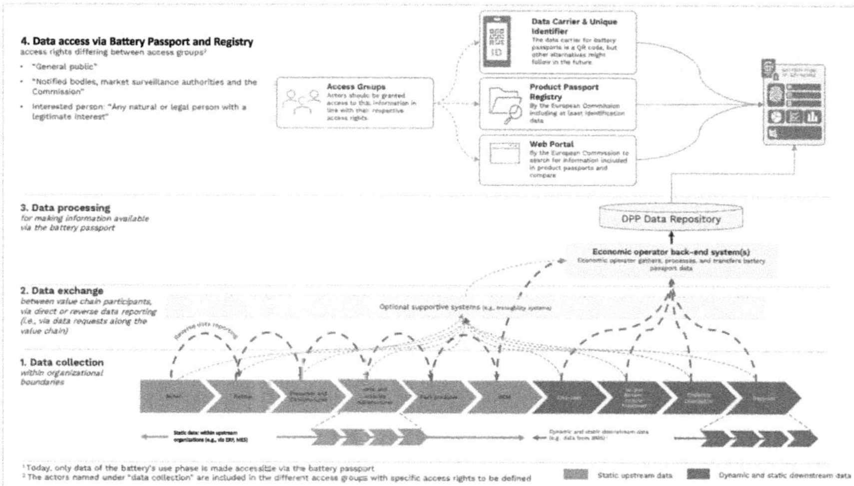

Figure 2.2 Technical design and operation of the battery passport. *Source:* The Battery Passport Content Guidance / Battery Pass Consortium.

require compatible infrastructure, which not all stakeholders currently have. As Battery Passports scale, this trade-off between simplicity and automation will likely influence how the system evolves and whether a hybrid approach becomes the industry norm (Table 2.3).

2.1.3 Real-Time Monitoring Systems

Real-time monitoring systems would be another essential component, designed to provide continuous updates on battery health and performance. These systems could empower stakeholders with actionable insights, enabling predictive maintenance and reducing the likelihood of failures. For example, by analyzing data on usage patterns and state-of-health metrics, manufacturers and operators could intervene proactively to address potential issues, optimizing the battery's efficiency and lifespan [1].

Real-time systems operate on a continuous feedback loop. Through constant measurement and edge-level processing, they alert operators to abnormalities like temperature spikes or voltage drops before they trigger system-wide issues. These insights are particularly valuable in high-utilization environments, such as electric vehicle (EV) fleets or stationary storage farms, where battery performance and reliability are mission-critical (Table 2.4).

However, it is important to acknowledge that fully integrated real-time monitoring may not materialize in the initial rollout of Battery Passports, if at all. Several barriers stand in the way. Technologically, real-time systems require advanced sensors, continuous connectivity, and robust data infrastructure, all of which add cost and complexity. Even less ambitious versions providing regular but noncontinuous updates may face similar roadblocks. Privacy concerns and data protection laws like GDPR raise red flags

Table 2.3 Evaluation of asset tracking solutions.

Feature	QR code	Barcode	RFID (active)	RFID (passive)	NFC	BLE	GPS
Cost-effective	☑	☑	✗	☑	✗	✗	✗
Real-time tracking	✗	✗	✗	✗	✗	☑	☑
Power consumption	✗	✗	☑	✗	☑	✗	☑
Range of scanning	High	Low	High	Low	Low	Low	Unlimited
Storage capacity	3 KB	>100 Bytes	2 KB	4–8 KB	48–8,000 B	NA	Unlimited
Continuous scanning	☑	☑	☑	☑	☑	At intervals	Real-time data
Two-way communication	✗	✗	☑	✗	☑	☑	☑
Labor intensive	☑	☑	☑	✗	☑	☑	☑
Popularity	Very high	Very high	High	High	Moderate	Moderate	Moderate

Source: Kashyap [27].

Table 2.4 Safety, performance, and environmental test methods for second-life battery evaluation.

Test	Test method description
OCV test	OCV must be measured at battery pack, system, module, and cell levels. Values must be compared and recorded. Reject if below secondary-use standards.
Incoming high-voltage insulation inspection	Measure insulation resistance between battery terminals and dead metal parts. Minimum 100 Ω/V for DC, 500 Ω/V for AC.
Capacity check	Charge fully, rest, then discharge using constant current/power and record capacity. Reject if below standard.
Internal resistance check	After a full charge and rest, discharge and measure voltage/current. Calculate $R = (V1-V2)/(I2-I1)$. Reject if above standard.
BMS control and protection inspection	Measure self-discharge rates over time. Reject if outside of standard.
Charge/discharge cycle test	Cycle the battery once at room temperature, and record all parameters. Reject if any exceed the standard.
Self-discharge inspection	Charge fully, rest, and measure OCV at intervals. Reject if value exceeds standard.

Table 2.4 (Continued)

Test	Test method description
Overcharge protection test	Charge beyond max voltage and perform full cycle and safety tests. Reject if failure or hazard occurs.
External short circuit test	Short the battery with a resistor, let it rest, then test. Reject if hazardous outcome.
Overdischarge protection test	Discharge to device protection limits, then test. Reject if safety is compromised.
Temperature and operation limit test	Cycle at max temperature limits multiple times. Reject if values exceed limits or a hazardous result occurs.
Unbalanced charging test	Charge with imbalanced cell/module conditions. Reject if safety standards are not met.
Cooling/heating system failure test	Simulate system failure and charge/discharge cycle. Reject if failure or hazard occurs.
Electrical components test	Test fan operation, leakage, and strain. Reject if burning or excessive strain deformation is observed.
Hot and cold cycle test	Cycle temperature extremes over 9 rounds. Reject if a hazardous result occurs.
Humidity resistance test	Perform a voltage test after IP-level humidity exposure. Reject if failure or hazard.
Salt spray test	Test according to standard. Reject if failure or hazard.
Internal flame exposure test	Thermal failure of a cell inside. Reject if combustion or explosion occurs.
External flame exposure test	Expose to external flame. Acceptable if no failure occurs.
Static test	Apply pressure on all sides. Reject if safety failure occurs.
Impact test	Drop the steel ball at different angles. Reject if safety failure occurs.
Drop test	Drop on a cement floor. Height depends on weight. Reject if failure occurs.
Molding stress test	Test at high temperature. Reject if a leak or break occurs.
Pressure relief valve test	Charge in mineral oil until the valve activates. Reject if there is no valve action or leakage.
Pressure relief valve open	Test 3 valves, and measure release start pressure. Pass if within 90–100% of nominal value.

Source: Adapted from Rufino Júnior et al. [3].

around how and where data is stored, especially if it includes user behavior or location. Intellectual property concerns further complicate matters, as manufacturers may be reluctant to expose sensitive performance data. Combined with the added financial and operational burden, these challenges mean real-time or even near-real-time monitoring features may be postponed indefinitely or remain confined to niche applications rather than becoming a core function of the Battery Passport.

2.1.4 Data Loggers and Edge Analytics

Data loggers play a crucial role in predictive maintenance by continuously monitoring and recording key operational metrics such as temperature, vibration, humidity, and pressure. These devices enable maintenance teams to identify patterns and detect anomalies that could lead to equipment failures. By providing continuous condition monitoring, data loggers help optimize maintenance schedules, reduce unplanned downtime, and extend equipment lifespan. Their ability to store and analyze historical data makes them essential tools for long-term performance assessment and proactive maintenance strategies [4].

Data loggers function both as historical archives and real-time diagnostic tools. Built to operate in rugged conditions, these devices can

- ***Record extensive datasets** including charge cycles, temperature fluctuations, and environmental stressors.*
- ***Enable predictive analytics** based on long-term trend recognition.*
- ***Support edge computing,** performing localized data processing before cloud upload to reduce latency and bandwidth use* [5].
- ***Survive harsh environments** such as high-vibration automotive applications* [6].

Data loggers are compliance instruments, ensuring batteries meet safety and durability standards under real-world conditions.

It is worth noting that many of these tools and capabilities already exist in most modern EVs. Vehicles are equipped with sophisticated monitoring systems that track battery health, performance, and safety metrics in real time. This connection facilitates decision-making, strengthens traceability, and paves the way for smarter, data-driven sustainability practices across the entire battery lifecycle. Rather than starting from scratch, the industry has a head start; the challenge is consolidating this data in a standardized and effective system.

Edge analytics refers to the processing of data at or near the source of generation, such as within a battery management system or onboard vehicle computer, rather than sending it all to a centralized server or cloud. For batteries, this means analyzing performance metrics like temperature, charge cycles, voltage fluctuations, and degradation patterns directly within the device or vehicle. The benefit is speed: critical insights can be acted on immediately, enabling faster responses to safety issues, optimizing energy use, or fine-tuning predictive maintenance. In the context of the Battery Passport, edge analytics would enhance the quality and timeliness of data fed into the digital record. Instead of waiting for periodic uploads or external diagnostics, the passport could reflect near-instant status updates, increasing its value for manufacturers, regulators, recyclers, and second-life users. It also reduces bandwidth costs and improves data security, as sensitive information does not have to be transmitted constantly or stored in a single vulnerable location.

2.1.5 Communication Infrastructure: Network Devices

Network devices bridge the gap between data collection and centralized storage by transmitting sensor data to cloud platforms. Depending on the deployment context, connectivity can rely on Wi-Fi, cellular (4G/5G), or satellite links, enabling near real-time access to operational data. Security and encryption depend on the chosen network and configuration (e.g., cellular security mechanisms defined by 3GPP, WPA3 for Wi-Fi, and encryption claimed by satellite providers) [7]. While these technologies may not appear in the first iterations of the Battery Passport, they are strong candidates for future integration as digital infrastructure matures and becomes more widely accessible. Features like real-time battery analytics, edge computing, and in-vehicle data synchronization could greatly enhance the passport's capabilities, bringing faster insights, deeper traceability, and more intelligent lifecycle management. As the ecosystem evolves, these advancements will likely become key components of a smarter, more connected sustainability platform.

This infrastructure ensures that no matter where a battery is located, inside an electric vehicle on the road or in a warehouse in another country, its data can be accessed, verified, and acted upon without delay. The backbone of the system must support

- Encrypted transmission protocols
- Global roaming capability (e.g., satellite uplinks for remote facilities)
- High-frequency, low-latency communication for edge artificial intelligence (AI) and instant alerts

While these communication technologies outline a compelling vision for seamless, real-time battery data exchange, it is important to recognize that much of this remains theoretical in the context of the Battery Passport. The infrastructure described (global cellular coverage, satellite uplinks, edge-AI processing) may not be feasible to implement in the near future. Technical feasibility, high deployment costs, data privacy laws like GDPR, and unresolved intellectual property concerns all pose significant barriers. Additionally, not all stakeholders across the value chain have the capacity or incentive to invest in high-bandwidth, always-on connectivity. The systems discussed here represent what could be possible if the Battery Passport evolves into a fully integrated, intelligent platform. But for now, they remain aspirational components: technologies that may enable future capabilities, not confirmed features of the initial rollout.

Moreover, the frequency of data communication, whether real-time, daily, weekly, or monthly, will significantly shape the choice of network infrastructure. Real-time systems may require cellular or satellite links with high-speed, low-latency capabilities, whereas less frequent updates could be handled through local Wi-Fi or even delayed batch uploads. Determining the right balance between data immediacy, cost, and technical complexity will be crucial in designing scalable, sustainable communication strategies for the Battery Passport system.

2.1.5.1 Cellular Communication

Cellular networks, such as 4G and 5G, play a critical role in enabling global data transmission, especially in remote or rural areas where other network infrastructures are unavailable. They are commonly used in EVs and industrial battery applications where frequent or real-time data transmission is essential. If the Battery

Passport evolves toward high-frequency updates, daily or real-time cellular connectivity may become a necessary backbone for mobile and high-value assets. However, such use would require balancing the benefits of continuous monitoring with the costs of data plans, infrastructure maintenance, and energy demands on devices.

While cellular networks provide extensive coverage and facilitate continuous monitoring, they come with higher costs associated with data plans, network setup, and ongoing maintenance. Additionally, network congestion in urban areas and limited availability in extremely remote regions may pose challenges. Nevertheless, the global reach and real-time connectivity offered by cellular networks make them indispensable for applications requiring constant monitoring and dynamic tracking of battery data [8].

2.1.5.2 Wi-Fi Communication

Wi-Fi technology is a popular choice for stationary battery systems, such as home energy storage units or fixed industrial batteries. With its high-speed data transfer capabilities, Wi-Fi is particularly effective in environments where local network infrastructure is readily available. It is cost-effective and relatively simple to implement, making it an attractive choice for settings where batteries remain fixed and proximity to a network is certain. In use cases where Battery Passport data is uploaded on a scheduled basis, such as weekly or monthly, Wi-Fi offers a practical solution that minimizes connectivity costs while still enabling timely updates, for example, at charging stations. The feasibility of Wi-Fi integration depends heavily on how often the system is expected to communicate and on the physical context in which the battery operates.

Despite its advantages, Wi-Fi is limited by its range and is more susceptible to interference, particularly in densely populated areas with overlapping networks. These limitations restrict its suitability to localized environments, but for stationary applications, Wi-Fi offers a fast, efficient, and economical solution for battery data transmission [8].

2.1.5.3 Bluetooth Communication

For short-range applications, Bluetooth technologyprovides an energy-efficient communication option. It is commonly deployed in small-scale battery systems, such as wearables, personal electronics, and home energy units. Bluetooth's low power consumption makes it ideal for devices that operate in energy-constrained environments. Additionally, its simple setup and compatibility with consumer electronics and IoT devices, often connecting through the user's phone or charging station, make it a convenient channel for periodic or event-triggered updates. If the Battery Passport is structured around less frequent or user-initiated data uploads, Bluetooth could serve as a lightweight solution for local sync and compliance.

However, Bluetooth's limited communication range and lower data transfer speeds restrict its use to scenarios where only small amounts of data need to be transmitted over short distances. While unsuitable for large-scale industrial or remote applications, Bluetooth remains an effective choice for low-power consumer electronics and localized battery systems [8].

Table 2.5 Evaluation of communication technologies.

Technology	Range	Speed	Energy efficiency	Cost	Applications
Cellular (4G/5G)	Long	High	Moderate	High	Remote battery systems, electric vehicles
Wi-Fi	Medium	High	Moderate	Low to medium	Stationary batteries, home energy systems
Bluetooth	Short	Low to medium	High	Low	Wearables, low-power consumer electronics

2.1.5.4 Comparison of Communication Technologies

Table 2.5 compares the key features of the communication technologies employed in the Battery Passport system.

2.1.5.5 Cloud Storage and Data Accessibility

Given the volume and complexity of data expected from the Battery Passport, cloud computing would play a vital role in managing and storing information if the data is shared in real time on a regular basis. Cloud platforms would offer scalability and remote accessibility, enabling seamless collaboration among manufacturers, recyclers, and regulatory bodies. By centralizing data, the system would ensure stakeholders can retrieve and analyze up-to-date information, supporting transparency and efficient decision-making [1].

In practical terms, cloud architecture allows for

- *Distributed access with role-based controls*
- *Real-time updates across global supply chains*
- *Integration with regulatory dashboards and machine learning (ML) tools*
- *Secure, long-term archiving for performance auditing and reporting* (Figure 2.3 and Table 2.6)

Although the Battery Passport may appear to be a single document, it is in fact a coordinated system of interoperable hardware and software tools. Each element, from embedded sensors to encrypted blockchains, ensures that batteries are not only powerful but also accountable. In the next section, we will explore how these technologies support data lifecycle management, covering how data is collected, validated, stored, and turned into insight.

Installation Processes
Installing the equipment for the Battery Passport system requires a well-planned approach, depending on whether it is being implemented for new batteries in a factory setting or retrofitted into existing battery systems. Below is a detailed guide on the installation processes for each scenario.

(Continued)

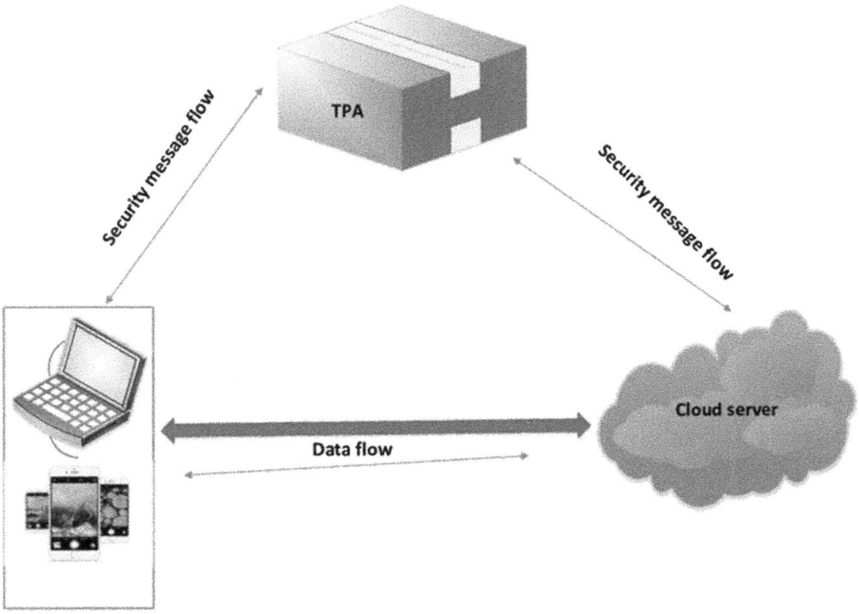

Figure 2.3 How cloud storage works. The architecture of the cloud storage service. *Source:* An efficient privacy preserving public auditing mechanism for secure cloud storage – Scientific Figure on ResearchGate. Available from: https://www.researchgate.net/figure/The-architecture-of-the-cloud-storage-service_fig1_316250675 (accessed 5 February 2026).

Table 2.6 Summary table: technology–function–stakeholder matrix.

Technology	Primary function	Key stakeholders benefiting
Smart sensors/IoT	Real-time data capture of battery health/performance	Manufacturers, fleet managers, EV operators
RFID tags	Remote lifecycle traceability and logistics	Recyclers, logistics operators, regulators
QR codes	Instant public access to lifecycle data	Consumers, service technicians, inspectors
Real-time monitoring	Continuous diagnostics and alerts	Operators, insurers, battery manufacturers
Data loggers	Long-term tracking, predictive maintenance, edge analytics	Engineers, OEMs, Environmental, Social and Governance (ESG) auditors
Network devices	Data transmission across regions	Global supply chains, data analysts
Cloud storage	Centralized data archive and analytics platform	Policymakers, industry consortia, platform providers
Blockchain and encryption	Data immutability, verification, secure transmission	Regulators, legal teams, all value chain actors

In-Factory Installation for New Batteries
When installing Battery Passport components for new batteries during the manufacturing process, the installation can be seamlessly integrated into the production line. This involves embedding sensors, data loggers, and network devices directly into the battery systems as they are assembled.
Steps:

- **Step 1: Sensor Integration:** *Sensors are installed into the battery modules to monitor parameters like temperature, voltage, and current. These sensors are placed in optimal locations within the battery to ensure accurate data capture without interference from other components* [9]. *This step is already in place for most batteries.*
- **Step 2: Data Logger Connection:** *Data loggers are installed to record the data provided by sensors. They are linked to the sensors via secure connections to ensure real-time data capture and logging* [9].
- **(unsure) Step 3: Network Device Setup:** *Network devices (Wi-Fi, cellular, or Bluetooth) are integrated into the battery system to transmit the collected data to cloud storage. These devices are configured to connect with the appropriate network based on the battery's use case (e.g., electric vehicles and energy storage systems)* [9, 10].
- **Step 4: System Calibration and Testing:** *Once installed, the system is calibrated to ensure that all components are working properly. Data transmission is tested, and initial diagnostics are run to validate sensor readings and data integrity* [9].

2.2 Data Lifecycle Management

The Battery Passport represents a stream of high-integrity data that follows a battery from production to end-of-life. Understanding how that data is collected, validated, stored, and analyzed is essential for grasping the full technical architecture of the system. This section breaks down the key stages in the data lifecycle and the technologies and safeguards that support it.

2.2.1 Types of Data Collected

The foundation of the Battery Passport is comprehensive, structured data gathered throughout a battery's entire lifecycle. This information spans physical characteristics, environmental metrics, performance statistics, and ethical sourcing disclosures, all unified in a standardized, digital format.

To ensure safety, environmental integrity, and global compatibility, the envisioned system would track compliance with safety and environmental standards. By documenting adherence to international certification requirements, such as ISO standards or regional sustainability mandates, the Battery Passport could simplify regulatory checks and reinforce consumer trust. This data would also serve as a vital tool for manufacturers and suppliers aiming to demonstrate their commitment to sustainable practices in an increasingly eco-conscious global market [1].

A critical category of data focuses on the carbon footprint and resource efficiency associated with batteries. By tracking emissions and resource utilization at every stage from raw material extraction to end-of-life disposal, the system could provide stakeholders with actionable insights. For example, this data might highlight areas where production processes could be optimized to reduce emissions or improve resource efficiency, helping the industry meet stringent environmental targets while fostering innovation in sustainable production techniques [1].

Transparency in the supply chain is a cornerstone of ethical production, particularly for materials such as lithium and cobalt that are often associated with environmental and social challenges. The Battery Passport would aim to provide a comprehensive view of material origins, ensuring that raw materials are sourced responsibly and in alignment with both human rights and environmental standards. This data would not only help stakeholders meet compliance obligations but also strengthen the industry's ethical alignment, addressing growing consumer and regulatory demands for greater accountability [1] (Figure 2.4).

During a battery's operational phase, real-time performance data such as charge cycles, temperature fluctuations, and overall health metrics could be continuously monitored and potentially one day be sent to the Battery Passport. This information could enable predictive maintenance, ensuring that batteries are used safely and efficiently while minimizing operational disruptions. Whether applied to electric vehicles, renewable energy storage systems, or consumer electronics, this data would be invaluable in optimizing battery performance and extending its useful life [1].

As batteries approach end-of-life, data becomes a driver of circularity. Accurate tracking of recyclable components, valuable materials, and recovery potential empowers recyclers and manufacturers to close material loops efficiently. By identifying valuable materials that can be recovered, such as lithium, nickel, and cobalt, the system could facilitate more efficient recycling processes and minimize waste. This approach would help manufacturers and recyclers close the loop on material usage, promoting responsible disposal practices and reducing environmental impact [1] (Figure 2.5).

Building on the concept of DPPs highlighted by Circularise, the Battery Passport would integrate structured, machine-readable data to ensure seamless accessibility for stakeholders [11]. *Information such as product identification, material composition, and lifecycle data would be unified under globally recognized standards, enhancing traceability and compliance. Moreover, mechanisms to balance transparency with data protection, such as anonymized sharing and encryption, would safeguard proprietary information while maintaining accountability* [11].

As this data is collected, ideally in real time and across supply chains, maintaining its accuracy and integrity becomes a challenge. The next section outlines how the system protects data from error, inconsistency, and tampering (Table 2.7 and Figures 2.6 and 2.7).

2.2.2 System Functioning

The Battery Passport system is designed to enable seamless and secure data flow to the cloud during production, and ideally throughout the battery's lifecycle as

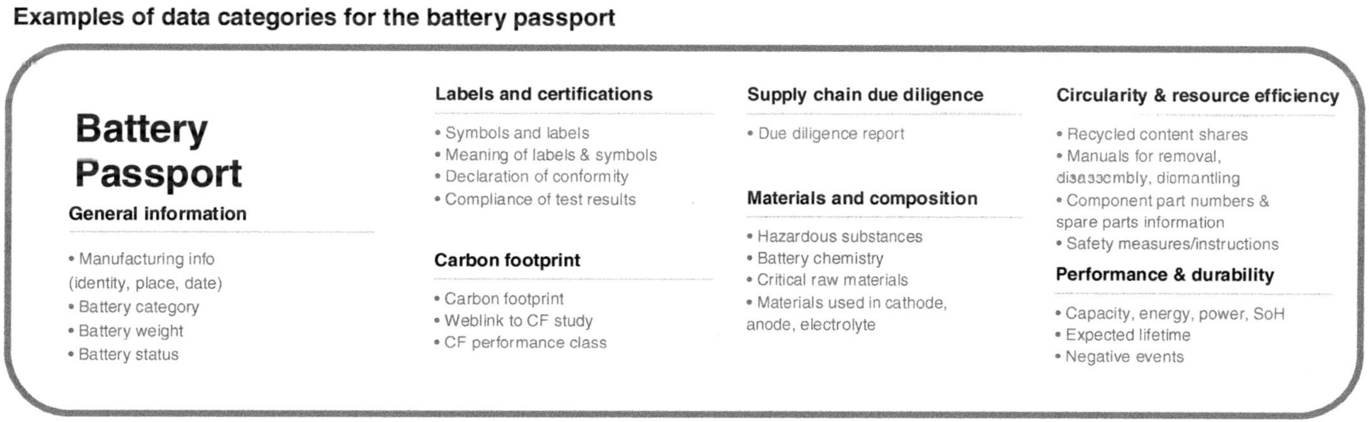

Figure 2.4 Data categories for the battery passport. *Source:* Battery Pass Consortium / https://thebatterypass.eu/assets/images/value-assessment/pdf/2024_BatteryPassport_Value_Assessment_Synthesis.pdf (accessed 15 December 2026).

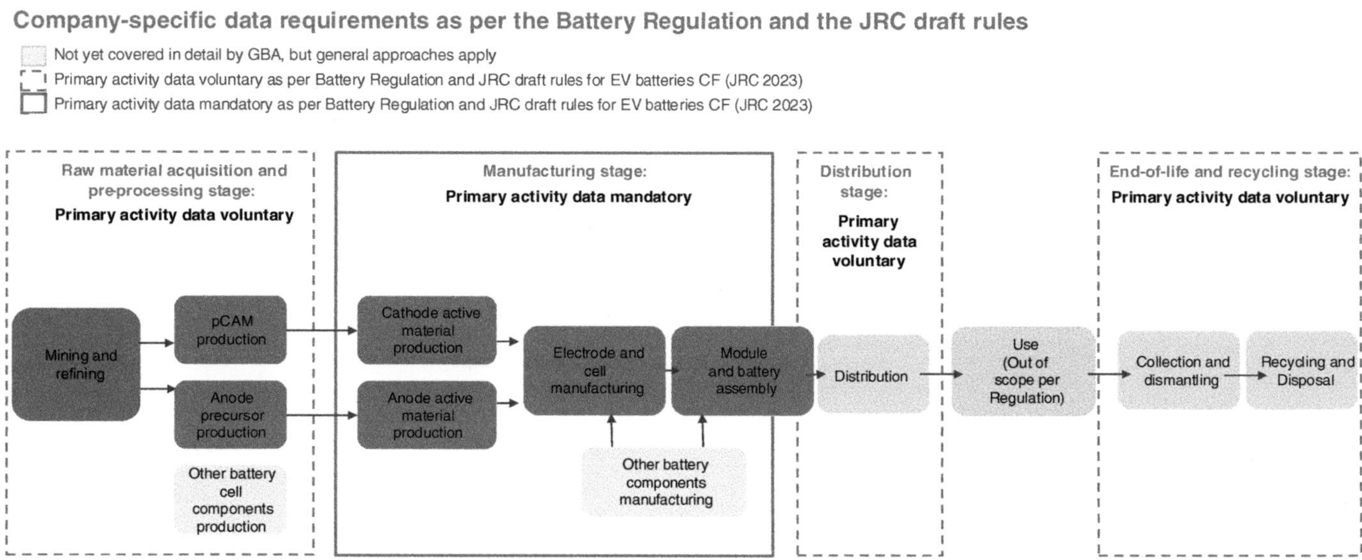

Figure 2.5 Enabled visualization via data aggregation thanks to the battery passport. *Source:* Battery Pass Consortium / Unlocking the value of the EU Battery Passport. https://thebatterypass.eu/assets/images/value-assessment/pdf/2024_BatteryPassport_Value_Assessment_Synthesis.pdf (accessed 12 November 2025) / CC BY-NC 4.0.

Table 2.7 Data collected during the initial screening process.

Areas for data collection	Collection of data
Battery system	Manufacturer details along with production date, capacity, specifications (including structure and type), and instructions. Further information in relation to user manual, cooling system, outage date and storage time, failure, abnormality, and maintenance records [3]
Module	Contains relevant information related to battery systems, such as manufacturer details, production date, capacity, component number, specifications, charge and discharge parameters, user manual, and structure and type [3]
Batteries	Like battery system and module, information such as manufacturer details, production date, weight and size, component number and marking, specifications, rated voltage and capacity, charge and discharge parameters, safety test data, expiration date, and internal structure (refer to IEEE 1625 or IEEE 1725) [3]
BMS	Manufacturer details, production date, component number, specifications concerning current, voltage and temperature protection, transmission protocol, CAN bus signal architecture, electrical circuit diagram, components list and configuration, and calculation logic [3]
Cooling system	Manufacturer details, production date, part number, specifications in relation to temperature, control, and flow, refrigerant substance: installation, operation, fault inspection, and maintenance [3]
Other systems	Manufacturer details, production date, component number, and specifications such as power capacity, dimensions, and usage parameters: installation, operation information on operation, troubleshooting, and maintenance records [3]

Source: Adapted from Rufino Júnior et al. [3].

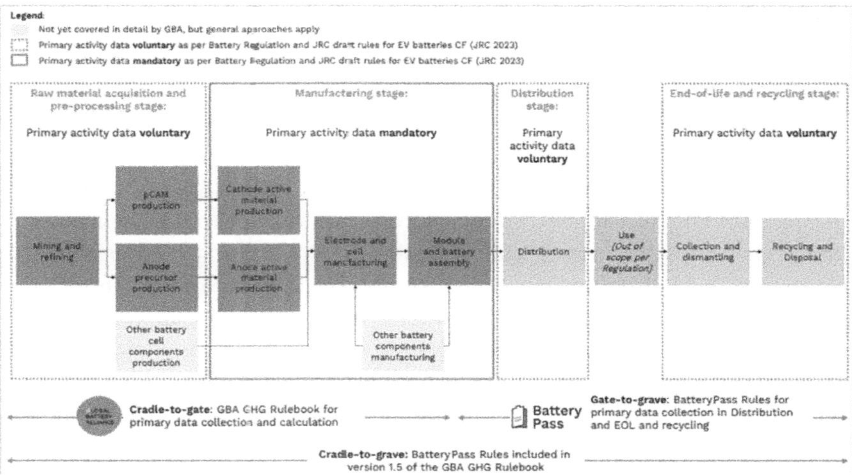

Figure 2.6 Company-specific data requirements as per the battery regulation and the JRC draft rules. *Source:* Battery Passport Content Guidance / Battery Pass consortium.

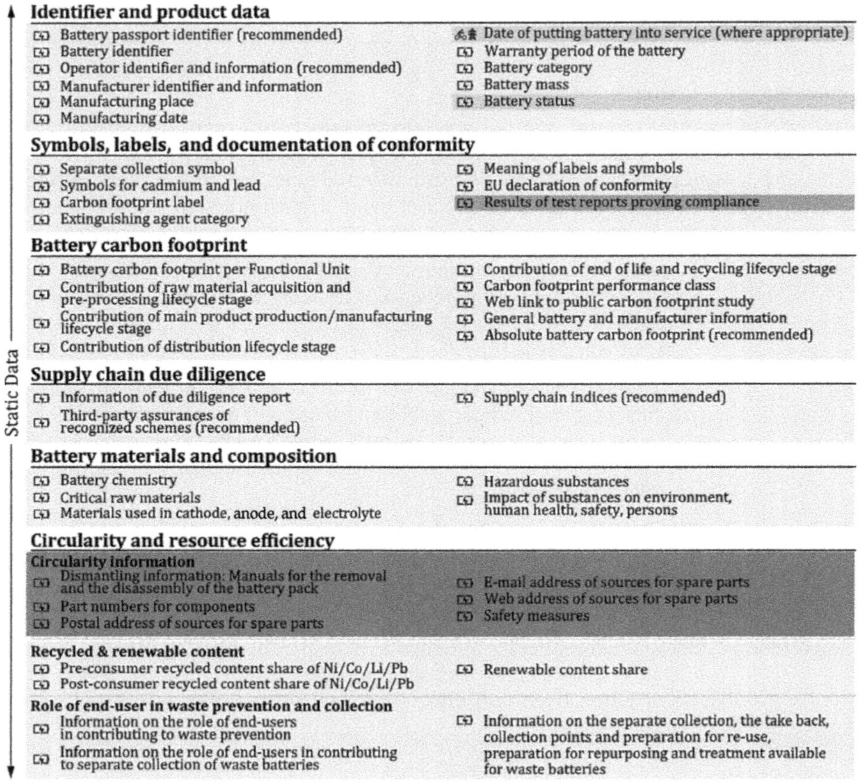

Figure 2.7 Data attributes overview. *Source:* Adapted from DIN DKE SPEC 99100.

well. This interconnected framework ensures that stakeholders can monitor, analyze, and report on battery lifecycle data, supporting transparency, compliance, and decision-making. We will discuss here how it will function (Figure 2.8).

2.2.3 Data Accuracy and Reliability

High-quality, credible data is the cornerstone of trust, regulation, and performance. The Battery Passport system uses layered validation technologies and advanced monitoring tools to ensure that every data point is both verifiable and dependable.

Automated data validation tools would cross-check all incoming data against predefined standards and benchmarks. These systems could detect inconsistencies or inaccuracies in real time, flagging issues for correction before they compromise the dataset. For example, data about material composition or carbon emissions could be compared to regulatory requirements, ensuring immediate compliance. Any deviations in performance metrics, such as battery capacity or charge cycles, would be identified and resolved promptly [1].

Validation is paired with systematic review processes that detect and resolve anomalies in the dataset, ensuring issues are identified before they cascade downstream.

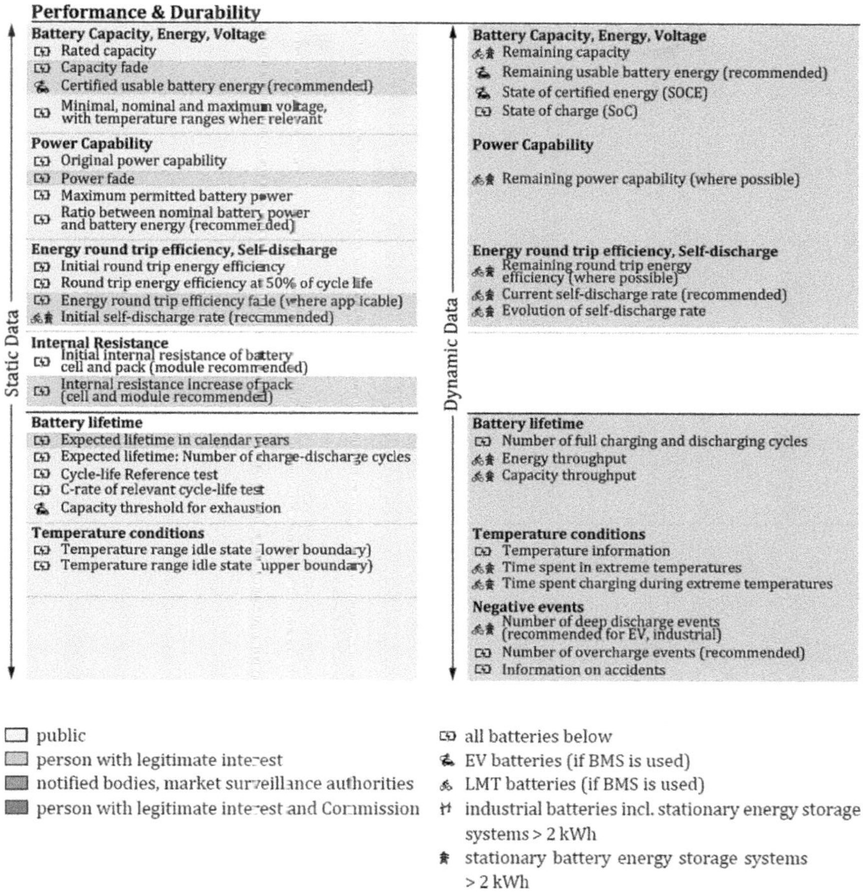

Figure 2.7 *(Continued)*

Regular quality assurance (QA) processes would involve systematic reviews of the data to identify missing entries, duplications, or errors. Paired with advanced anomaly detection systems, these QA measures would highlight unusual patterns, such as unexplained spikes in emissions data or discrepancies in material sourcing records, and trigger automated alerts for further investigation, allowing stakeholders to address potential issues proactively [1].

With collection and accuracy secured, the next challenge is scale how to store billions of data points in a way that is reliable, accessible, and secure.

Data Flow and Component Integration

At the heart of the system architecture is the flow of data through four key stages: collection, logging, transmission, and storage. Sensors, data loggers, network devices, and cloud services work in unison to create a streamlined process that ensures data accuracy and accessibility.

(Continued)

Overview of update-frequency requirements for dynamic data attributes

Performance data attributes

Subtopic	In-use/ Dynamic Data
Capacity, Energy, SoH, voltage	- 1 Capacity fade (in %)[a] - 2, 5 Remaining capacity (voluntary for all batteries)[b] - 3 Remaining usable battery energy (voluntary)[b] - 3 State of certified energy (SOCE)[b] - 1 State of charge (SoC)[b]
Power Capability	- 1 Power fade[a] - 2, 5 Where possible, Remaining power capability[b]
Energy round trip efficiency, Selfdischarge	- 1 Where applicable, round trip energy efficiency fade[a] - 2, 5 Current selfdischarging rate[b] - 2, 5 Evolution of self-discharging rate[b]
Internal Resistance	- 1 Current internal resistance (pack; voluntary: cell/module)[a] - 1 Internal resistance increase (pack; voluntary: cell/module)[a]

Durability data attributes

Subtopic	In-use/ Dynamic Data
Expected Lifetime	- 1 Current number of (full) charging and discharging cycles[b] - 2, 5 Capacity throughput[b] - 2, 5 Energy throughput[b]
Temperature conditions	- 1 Time spent charging during extreme temperatures[b] - 1 Time spent in extreme temperatures[b]
Negative events	- 2, 5 Number of deep discharge events[b] (voluntary for EV, industrial) - 1 Number of overcharge events (voluntary)[b] - 1 Accidents[b]

Access groups

- Public
- Persons with legitimate interest

Battery categories

1 All batteries in battery passport scope
2 LMT batteries (if BMS is used)
3 EV batteries (if BMS is used)
4 Industrial batteries[1] incl. stationary BESS > 2 kWh
5 Stationary BESS > 2 kWh (subcategory to industrial)

Update requirements

a) From Art. 10; to be made available upon placement on market and status change (Re-use/repurpose...)
b) Dynamic reporting depending on definition of 'up-to-date' in Art 77/Annex XIII or Art. 14

1) Industrial batteries mentioned for completeness, no data attribute assigned separately

Figure 2.8 Battery storage in power system. *Source:* Batteries and Secure Energy Transitions (IEA) / IEA / CC BY 4.0.

Data Collection: *Sensors embedded in the battery collect real-time data on performance and safety metrics, such as SoC and SoH. These insights are vital for tracking battery health and identifying potential issues early.*

Data Logging: *The collected data is organized and temporarily stored in data loggers, preserving its integrity until transmission to the cloud. This step ensures resilience against network disruptions.*

(Unsure) Data Transmission: *Network devices securely transfer data from the battery to the cloud using encrypted communication protocols, ensuring privacy and reducing the risk of data breaches.*

Data Storage and Analysis: *Cloud platforms store the transmitted data, providing scalability and security. Advanced tools within the cloud allow stakeholders to analyze data, derive insights, and ensure compliance with regulatory standards.*

GS1 Architecture for Digital Product Passports

GS1 standards provide a comprehensive architecture for DPPs, establishing robust foundations for product lifecycle traceability, including for products such as batteries. This system is designed to support the European Union's regulatory frameworks on sustainable products by providing interoperable and scalable solutions.

Persistent Global Identification

Central to GS1's architecture is the persistent global identification of products through the Global Trade Item Number (GTIN). This standard assigns each product or batch a unique and enduring identifier, ensuring consistent traceability from manufacturing through distribution, use, and end-of-life management. These identifiers are encoded using barcodes or RFID tags directly on the product or its components, ensuring durability and continuous access throughout the product lifecycle.

Decentralized and Open Architecture

GS1 emphasizes a decentralized approach, avoiding reliance on central databases or single-service providers. Product data is accessed through decentralized resolvers connected directly to the product's unique identifier via the GS1 Digital Link standard. This avoids single points of failure and reduces dependence on specific vendors or service providers. The decentralized structure fosters a flexible ecosystem, capable of integrating various stakeholders and data sources seamlessly.

Event-based Traceability with EPCIS

Lifecycle events—such as production, shipping, usage, repair, and recycling—are recorded systematically using the Electronic Product Code Information Services (EPCIS) standard. EPCIS facilitates structured data exchange among stakeholders, enabling seamless interoperability across the supply network. This system ensures transparency throughout the product lifecycle, supporting effective lifecycle management and regulatory compliance.

Secure and Flexible Data Access via GS1 Digital Link

(Continued)

(Continued)

The GS1 Digital Link standard integrates unique product identifiers into web-based URLs or QR codes, allowing direct and secure access to comprehensive product information. Stakeholders can retrieve product-related data easily through standard web technologies without requiring proprietary software. This model enables various users—from consumers and recyclers to regulators and manufacturers—to access tailored information relevant to their specific needs and contexts.

Scalability and Future-proofing
GS1's open standard-based approach ensures flexibility and adaptability to evolving technological and regulatory landscapes. By leveraging globally recognized, interoperable standards, the GS1 architecture readily accommodates new types of data, emerging technologies, and future regulatory requirements, maintaining its effectiveness over the long term.

Source: https://gs1.eu/wp-content/uploads/2022/08/Digital_Product_Passport_Architecture_GS1inEurope_March_2022.pdf.

2.2.4 Data Storage Solutions

Capturing accurate data is only part of the story. The Battery Passport must also store this data in ways that support cross-border collaboration, real-time access, and long-term digital preservation. Cloud infrastructure answers that need.

Cloud storage enables seamless data access and updates from anywhere, fostering collaboration across industries. By allowing stakeholders to retrieve and modify data in real-time, cloud platforms support remote decision-making and coordination across value chains [12].

Given the sensitive nature of stored data, cloud providers implement strong security measures, including encryption, role-based access control, and multifactor authentication, to protect information integrity and prevent unauthorized access [12].

Auto ID (Automatic Identification)

Auto ID technologies such as 2-D matrix codes and RFID tags play a crucial role in tracking batteries and their components. By tagging batteries with these identifiers:

Batteries can be easily identified at every stage of their lifecycle, from manufacturing to recycling.

Each transfer of the battery be it from manufacturers to distributors, or from consumers to recyclers is documented, ensuring full traceability. For example, an RFID tag attached to a battery during production could later provide recyclers with valuable data on material composition, aiding efficient recycling processes. Auto ID thus forms the backbone of physical-to-digital data integration in the Battery Passport system.

The effective operation of the envisioned Battery Passport system requires scalable and secure data storage solutions to manage extensive and dynamic information across the battery lifecycle. As the system functions as a digital twin for each battery, robust infrastructure is essential for capturing and preserving data related to production, usage, and end-of-life processes. Given the increasing volume of battery-related data, storage solutions must prioritize efficiency and adaptability. DPPs have been explored with various storage architectures, combining decentralized and cloud-based frameworks to ensure data accessibility and security while mitigating risks associated with centralized database [13].

Once the data is collected, validated, and stored, it becomes a tool for action. Through analysis, this data turns from static record to strategic asset.

However, it is important to clarify that not all data collected across a battery's lifecycle will necessarily be stored in the Battery Passport itself or even in the cloud. In many cases, especially for technical parameters like charge cycles, temperature events, or degradation trends, data may be stored internally within the battery's own management system. These embedded storage systems are designed for real-time performance and localized diagnostics, enabling immediate actions like performance adjustments or safety shutdowns without relying on external connectivity. Even if the Battery Passport includes a digital record of key lifecycle events, it may only reflect selected or aggregated data while the more detailed logs remain locked inside the battery.

As a result, the Battery Passport may function less like a full digital twin and more like a curated snapshot offering transparency where needed for compliance, recycling, or resale, but not disclosing the full scope of telemetry captured by the battery system. Whether this model shifts toward deeper integration over time will depend on evolving standards, trust frameworks, and regulatory requirements around data ownership, access, and interoperability.

In other words, at this stage, it is still unclear how much of the battery's operational data will actually be included in the Battery Passport and how much will remain stored internally or managed through separate systems. The final structure will depend on evolving regulations, industry consensus, and practical trade-offs between transparency, privacy, and technical feasibility. We do not yet know whether the passport will act as a full digital twin, a simplified summary, or something in between. What is certain is that different models are being tested, and decisions around what data gets shared, where it is stored, and who can access it will significantly shape how the Battery Passport works in practice.

Artificial Intelligence (AI)

AI algorithms enhance the utility of the data collected by tracking technologies by providing actionable insights. These algorithms

Analyze performance data to predict a battery's end-of-life, enabling stakeholders to plan for recycling or second-life applications proactively.

(Continued)

(Continued)

Identify patterns and anomalies in battery usage, optimizing maintenance schedules and extending battery lifespans. For example, AI might detect declining performance in an electric vehicle battery and recommend preemptive recycling, preventing potential failure and ensuring resource efficiency. By integrating AI, the Battery Passport system can make data-driven decisions that benefit both stakeholders and the environment [1].

2.2.5 Data Analysis Techniques

Analyzing lifecycle data unlocks the full potential of the Battery Passport system. Throughout a battery's lifecycle, substantial amounts of data are generated, and effectively analyzing this information enables stakeholders to optimize battery performance, operational processes, and sustainability outcomes.

Leveraging Big Data

Big data serves as a powerful enabler within the Battery Passport system, providing comprehensive insights that help stakeholders across the battery lifecycle from design to recycling make informed decisions that drive sustainability and operational efficiency. By analyzing large datasets generated throughout the lifecycle, stakeholders can enhance battery performance, extend lifespan, and improve resource recovery, all while promoting transparency and reducing environmental impact.

Enhancing Battery Design
Big data plays a crucial role in battery manufacturing, enabling manufacturers to analyze extensive datasets on performance prediction, defect detection, and process optimization. By leveraging data-driven insights, companies can enhance production quality, reduce costs, and improve the efficiency of lithium-ion batteries across diverse applications, such as EVs and renewable energy storage systems [14].

Optimizing Battery Usage
During the usage phase, big data plays a crucial role in providing real-time insights into how batteries perform under different operating conditions and user behaviors. By analyzing usage patterns, manufacturers and users can identify optimal charging cycles, prevent overcharging or deep discharges, and predict maintenance needs. This extends the lifespan of batteries and improves overall efficiency, minimizing the frequency of replacements and reducing waste [8].

Improving Recycling Efficiency
Big data is equally critical in optimizing recycling processes. By providing detailed information about a battery's material composition, manufacturing methods, and usage history, big data enables recyclers to fine-tune recovery processes, maximizing

the extraction of valuable materials like lithium, cobalt, and nickel. This not only enhances resource recovery but also supports circular economy practices by reducing the environmental impact of battery production and disposal [15].

Enabling Transparency and Sustainability
The Battery Passport system relies on big data to bring unprecedented transparency to the global battery supply chain. By aggregating trusted data on sourcing, production, usage, and recycling, the system allows stakeholders to design more sustainable production and recycling processes. This transparency ensures that batteries contribute to circular economic practices and support broader environmental goals, such as reducing carbon emissions and promoting responsible sourcing [9].

Enhancing Supply Chain Management
Big data also enables manufacturers to optimize supply chain management by providing real-time monitoring of material sourcing, logistics, and production processes. Through predictive analytics and real-time data tracking, stakeholders can identify inefficiencies, reduce material waste, and prevent supply chain disruptions. Additionally, by aligning supply chain activities with sustainability goals, manufacturers can minimize the environmental impact of battery production while ensuring operational efficiency [10].

Big data is integral to the success of the Battery Passport system. It provides the foundation for optimizing battery design, improving usage patterns, and enhancing recycling efficiency, while also promoting transparency and sustainability across the entire battery lifecycle. By leveraging big data, stakeholders can make better decisions that support the long-term viability of batteries in a circular economy.

Descriptive analysis forms the foundational layer by summarizing historical data to clarify what has occurred. It identifies key metrics and usage patterns, providing stakeholders with a comprehensive snapshot of battery conditions and performance. Manufacturers can leverage this analysis to evaluate product lines, while recyclers can assess the effectiveness of their recovery processes, establishing an essential baseline for further analysis.

Diagnostic analysis expands upon descriptive insights, exploring the reasons behind observed trends or anomalies. By closely examining detailed performance data, this method identifies root causes of issues such as uneven degradation rates or production inefficiencies. It supports manufacturers in pinpointing production bottlenecks and assists users in understanding battery performance inconsistencies under varying conditions, fostering targeted improvements.

Predictive analysis introduces forward-looking capabilities, utilizing historical data and statistical modeling to anticipate future scenarios. This approach helps stakeholders foresee potential battery failures, lifespan estimates, and recycling requirements. Manufacturers and users can proactively manage maintenance and resource allocation, minimizing disruptions and enhancing operational efficiency.

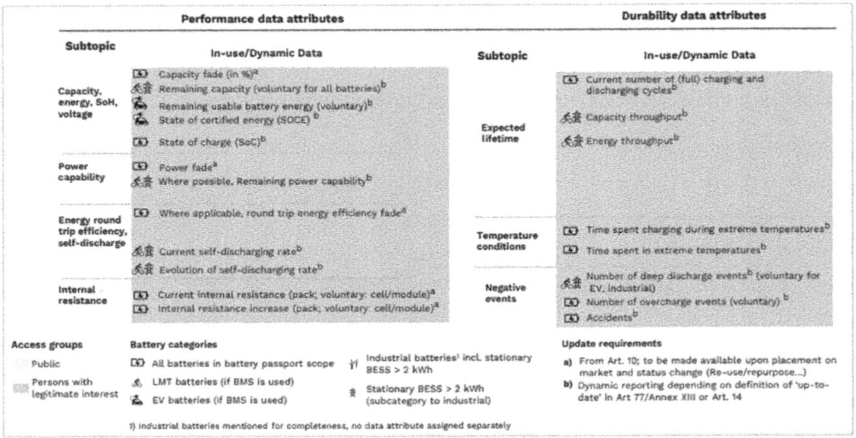

Figure 2.9 Overview on up-to-date requirements for dynamic data attributes. *Source:* Battery Passport Content Guidance / Battery Pass consortium.

For example, predictive models can forecast service intervals, enabling preventive actions that extend battery life and sustainability.

Building on predictive insights, prescriptive analysis goes further by offering actionable recommendations for optimal outcomes. Employing advanced optimization algorithms, it guides manufacturers toward improved durability techniques and advises recyclers on refining material recovery processes. This analysis facilitates informed, proactive decision-making, driving enhanced performance and sustainability. ML significantly enhances these analytical techniques by continuously refining insights as new data becomes available. Its adaptive nature excels in detecting subtle patterns that traditional methods might overlook, thus dynamically improving manufacturing strategies and recycling methods. By integrating ML, the Battery Passport system maintains responsiveness to evolving stakeholder needs and technological advancements, ensuring continual improvement and relevance (Figure 2.9).

Together, these data lifecycle processes collection, validation, storage, and analysis form the operational core of the Battery Passport. But with such powerful data flows come equally critical risks. The next section explores how the system secures this information against breaches, misuse, and manipulation ensuring that transparency is matched with trust.

2.3 Cybersecurity and Privacy

Safeguarding the vast amounts of data managed by the envisioned Battery Passport system is essential for maintaining stakeholder trust and ensuring the system's integrity. To achieve this, the system will employ a range of robust cybersecurity measures designed to protect data from unauthorized access, tampering, and breaches. These measures form a multilayered defense that secures the entire lifecycle of data within the system.

2.3.1 Blockchain for Privacy and Security

The integration of blockchain and cryptographic security mechanisms enhances both privacy and data integrity within the system:

> *Rather than choosing between transparency and privacy, blockchain systems allow for both. The key lies in selective disclosure and verifiable claims that do not compromise commercially sensitive data.*
>
> *By decentralizing control and integrating advanced cryptographic validation, the system mitigates the risks of centralized data breaches while enhancing trust across stakeholders* [16].
>
> *Security is reinforced by*
>
> - **End-to-end Encryption Protocols:** *Secure transmission between devices and cloud systems.*
> - **Public Key Infrastructure (PKI):** *Ensures only authenticated parties can alter or add data.*
> - **Digital Signatures:** *Confirm the origin of data entries, essential for compliance with sustainability reporting standards.*

- **Data Integrity and Trustworthiness:** The use of cryptographic certificates and Physical Unclonable Functions (PUFs) ensures that battery lifecycle data remains secure and resistant to tampering, providing a trustworthy foundation for supply chain transparency.
- **Controlled Access to Sensitive Information:** Secure certificate-based verification mechanisms allow selective data disclosure, ensuring that critical lifecycle and compliance records remain accessible only to authorized stakeholders while preventing unauthorized alterations.
- **Verifiable Traceability:** The application of cryptographic techniques enables transparency in supply chain transactions while maintaining confidentiality, ensuring that product lifecycle data can be verified without exposing unnecessary details.

This technology would address critical concerns around data manipulation and fraud, fostering confidence among all stakeholders while supporting sustainability and ethical sourcing initiatives [1].

Blockchain adds trust where traditional systems introduce doubt. Every transaction whether an ESG declaration or a battery's recycling report is time-stamped, cryptographically sealed, and immutable.

2.3.2 Encryption

This ensures that even if bad actors intercept or access stored data, its encrypted state renders it unreadable. Encryption is a key part of regulatory compliance, especially in jurisdictions with strong data protection laws.

In the context of the Battery Passport, encryption must be applied both in transit and at rest. That means protecting data as it moves between devices, cloud platforms, and end users, as well as when it is stored on servers or embedded systems. Transport Layer Security (TLS) and Advanced Encryption Standards (AES) are commonly used protocols to achieve this, ensuring that no single point of the data lifecycle becomes vulnerability (Figure 2.10).

Beyond technical implementation, encryption supports risk management by limiting liability exposure in the event of a breach. It also plays a critical role in access control – enabling granular permissions, where different stakeholders can access only the data they are authorized to see. Combined with audit logs and user authentication layers, encryption forms the backbone of a secure and transparent digital infrastructure for battery lifecycle data.

Moreover, as supply chains grow more complex and global, encryption supports cross-border data sharing while complying with regional laws like the EU's GDPR or California's California Consumer Privacy Act (CCPA). In this way, encryption both protects user privacy and enables the trusted data exchange that the Battery Passport relies on to function at scale.

2.3.3 Secure Data Transmission

To protect battery lifecycle data as it moves between systems, secure transmission channels are essential. All updates and exchanges must pass through encrypted communication protocols such as TLS, ensuring that the data remains confidential and tamper-proof from the moment it leaves the device to its destination, whether that is a cloud server, a regulatory platform, or another authorized system.

In a distributed and constantly updating environment, securing data in transit is as critical as securing it at rest. Industrial IoT systems are especially vulnerable

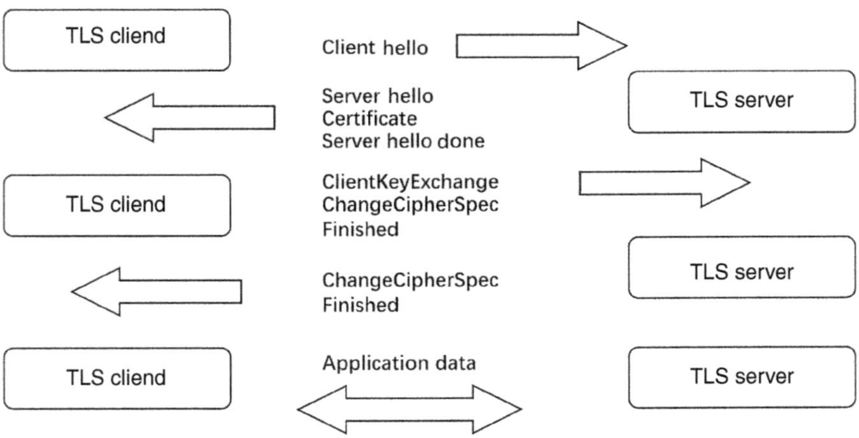

Figure 2.10 Transport layer security. *Source:* Ixela [17] / Medium.

to man-in-the-middle attacks, where malicious actors intercept or alter data as it moves between endpoints. By implementing strong encryption and enforcing endpoint authentication, the system can prevent unauthorized access and data manipulation.

Secure transmission also allows for integrity checks and real-time validation, helping detect any anomalies or unauthorized changes immediately. This not only maintains the reliability of the data but also builds trust among stakeholders ensuring that every actor in the value chain can rely on the accuracy and security of the information they receive.

Moreover, many data protection regulations now mandate encryption during transmission as a baseline requirement. As such, secure channels are a legal necessity for any Battery Passport system operating within or interacting with regulated markets.

2.3.4 Secure Storage Through Decentralization

To mitigate the risks inherent in centralized data systems, distributed storage architectures are increasingly being adopted. In such models, data is replicated and stored across multiple servers or nodes, which removes the dependence on any single system or location. This approach significantly reduces the risk of catastrophic failure due to server outages, cyberattacks, or infrastructure disruptions.

By distributing data across a network, the system eliminates single points of failure, one of the primary vulnerabilities in traditional centralized setups. If one node experiences a failure or is compromised, others continue to maintain access and integrity, ensuring that critical battery lifecycle data such as usage logs, recycling records, or compliance documentation remains continuously available and accurate.

Redundancy is key to this resilience. Even in the event of regional outages, localized cyber threats, or physical damage to infrastructure, mirrored data copies ensure continuity. This is particularly important in the context of regulatory compliance and operational traceability, where access to up-to-date records must be guaranteed under all circumstances.

Furthermore, distributed systems allow for more flexible access control and improved load balancing, supporting high-volume environments like those anticipated for the Battery Passport. As the battery ecosystem scales, so must the infrastructure that supports it and distributed storage provides the backbone for this future-ready design.

2.3.5 Identity and Access Management

Controlling access to sensitive battery data is fundamental to maintaining security and trust across the Battery Passport ecosystem. IAM systems can ensure that only authorized users such as manufacturers, recyclers, service providers, and regulatory bodies can view, modify, or upload data. These systems use multifactor authentication, digital certificates, and role-based access controls to tightly govern user permissions.

Rather than offering blanket access, the system assigns access rights based on user roles and responsibilities. For instance, a recycler may be granted access to end-of-life data, while a manufacturer can input production details but cannot edit compliance records. This segmentation of access reduces the attack surface and limits the risk of data leaks, misuse, or tampering.

Strong IAM also plays a critical role in mitigating both insider threats and external cyberattacks. By enforcing least-privilege principles where users only have the minimum level of access needed to perform their function, the system minimizes the impact of compromised credentials or malicious intent.

In high-compliance environments, IAM systems also support audit trails and traceability, making it easy to verify who accessed what data, when, and why. This safeguard, which enhances accountability, is increasingly required by regulatory frameworks governing DPPs and industrial data systems.

2.3.6 Consensus Mechanism

The integrity of the Battery Passport system can be further bolstered by its blockchain's consensus mechanism:

All transactions and updates must be validated by the network, ensuring that every entry is legitimate and agreed upon by participants.

This prevents unauthorized modifications and adds another layer of security to the system. For example, if a manufacturer submits data about a battery's material composition, the network must verify and approve the entry before it is permanently recorded [1].

Consensus mechanisms both validate transactions and create accountability. Each new entry has a clear origin and must pass through a standardized trust protocol, reducing errors and fraud.

Securing the system is only one part of the equation. Protecting the privacy of the data within it is equally critical.

2.3.7 Addressing Privacy Concerns

The envisioned Battery Passport system must navigate a delicate balance between transparency and data privacy. Given the sensitive nature of the data it manages, ranging from manufacturing details to lifecycle and recycling information, privacy concerns are central to its design and implementation. By adhering to stringent regulations like the GDPR and integrating privacy-preserving technologies, the system ensures data is handled responsibly while maintaining its core functionality (Figure 2.11).

2.3.8 Data Access Control

The envisioned Battery Passport system must navigate a delicate balance between transparency and data privacy. Given the sensitive nature of the data it manages, ranging from manufacturing details to lifecycle and recycling information,

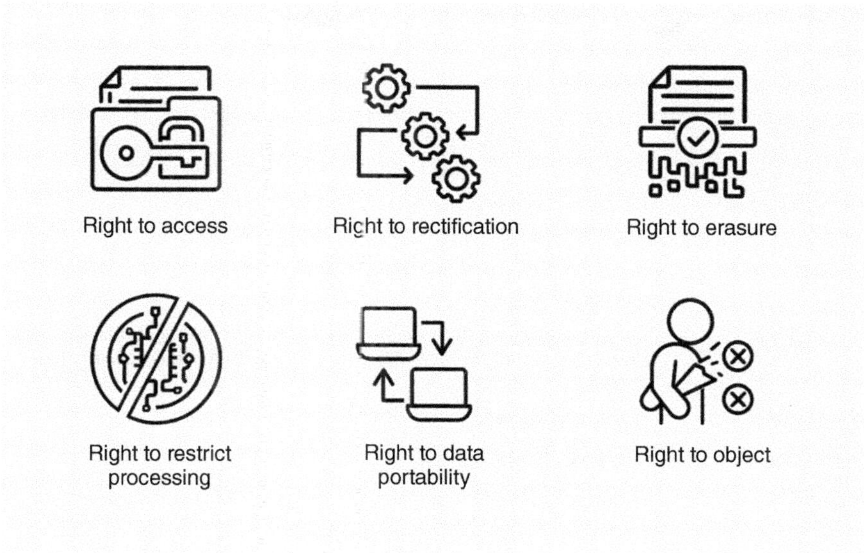

Figure 2.11 GDPR individual rights.

privacy concerns are central to its design and implementation. Privacy models in blockchain-based systems emphasize controlled disclosure of sensitive data through permissioned access and encryption mechanisms.

To align with GDPR standards, the Battery Passport system should incorporate structured mechanisms for lawful and secure data handling. Best practices include

- **Data Minimization:** Implementing permissioned blockchain technology to store only necessary transaction data while utilizing off-chain storage for bulk data [18].
- **Role-based Access:** Using smart contracts to control disclosure based on e-consent authorization, ensuring that access is granted only to predefined stakeholders [18].
- **Security and Encryption:** Blockchain-based access control frameworks employ encryption schemes and smart contract governance to ensure the integrity and confidentiality of stored data [18].

These practices help satisfy regulatory obligations while limiting the potential misuse of data. By ensuring data is used only for its intended purpose, the system reduces legal risk and reinforces trust among participants (Table 2.8).

2.3.9 The Role of Private Blockchains

Private blockchains play a critical role when heightened privacy is required (Table 2.9).

Private blockchains offer a secure and controlled environment for managing sensitive data within the battery value chain. Their restricted access ensures that only

Table 2.8 Key milestones in the evolution of EU data protection law.

Date	Event
24 October 1995	*Adoption of the European Data Protection Directive (Directive 95/46/EC), establishing foundational privacy rules in the EU.*
4 November 2010	*European Commission issues a Communication on "A comprehensive approach on personal data protection in the EU."*
25 January 2012	*European Commission proposes a comprehensive reform of the 1995 data protection rules.*
27 April 2016	*European Parliament and Council adopt the General Data Protection Regulation (GDPR) and the Law Enforcement Directive (LED).*
5 May 2016	*LED enters into force.*
24 May 2016	*GDPR enters into force.*
6 May 2018	*Deadline for Member States to transpose the LED into national law.*
25 May 2018	*GDPR becomes applicable across the EU.*
23 October 2018	*Adoption of the EUDPR (Data Protection Regulation for EU institutions).*
11 December 2018	*EUDPR becomes applicable.*
24 June 2020	*First Commission report on the application of the GDPR is published.*
25 July 2022	*First Commission report on the application of the LED is published.*
14 October 2022	*First report on the application of the EUDPR is published.*
4 July 2023	*Commission proposes further specification of procedural rules for GDPR enforcement.*
25 July 2024	*Second Commission report on GDPR application is published.*
21 May 2025	*Commission proposes targeted changes to record-keeping obligations.*

Table 2.9 Key data access and privacy control concepts.

Concept	Description
Restricted access	Only trusted parties, such as manufacturers, regulators, or recyclers, are granted access to specific data.
Granular permissioning	Proprietary or confidential data, such as detailed recycling metrics or supplier agreements, is shielded from unauthorized entities.
Purpose limitation	The use of private networks ensures data is processed only for its intended purpose, adhering to GDPR's principles of purpose limitation and data minimization.

authorized stakeholders, such as manufacturers, regulators, and recyclers, can interact with specific datasets. To enhance data security, granular permissioning mechanisms are often implemented, safeguarding proprietary data such as supplier agreements and recycling metrics. Additionally, private blockchain architectures adhere to purpose limitation principles by ensuring that data is processed exclusively for predefined regulatory or operational needs, reducing the risk of misuse. These features collectively support regulatory compliance and foster trust among stakeholders by balancing transparency with the protection of commercially sensitive information [12].

These features are especially important as the Battery Passport becomes a business-critical tool. Private blockchains help ensure companies can engage in data sharing without exposing trade secrets or violating legal obligations.

As privacy expectations and regulations evolve, the Battery Passport will need to incorporate secure data management practices, such as blockchain-based traceability and access controls, to maintain compliance with data protection regulations like GDPR. DPPs are already being designed with protective mechanisms that balance transparency with data security, ensuring responsible data governance in sustainable supply chains [13].

2.4 Interfaces and Accessibility

A robust technical system is only as effective as its interface allows it to be. For the Battery Passport to succeed across its diverse network of users including manufacturers, consumers, and recycling facilities, its interfaces must be intuitive, inclusive, and operable under real-world conditions. This section outlines how interface design and accessibility features can maximize adoption, ease of use, and equity within the system.

In complex, data-driven systems like the Battery Passport, the importance of interface design has less to do with aesthetics than with adoption and effective long-term implementation. If users cannot easily find, interpret, and act on the data, the system fails regardless of how powerful its backend may be.

2.4.1 Designing User-Friendly Interfaces

A well-designed user interface (UI) plays a pivotal role in ensuring that the Battery Passport system is accessible and efficient for a broad spectrum of stakeholders, including manufacturers, consumers, and recycling facilities. By offering an intuitive and user-friendly experience, the system can significantly enhance user engagement, streamline workflows, and promote better sustainability practices across the battery lifecycle. Below are the key reasons why user-friendly design is essential for different stakeholders.

2.4.1.1 Manufacturers

Manufacturers are responsible for managing vast amounts of data related to battery production, compliance, and performance. A user-friendly UI can help streamline their operations by offering intuitive access to data input, analysis tools, and compliance tracking systems. This ease of use not only reduces the likelihood of data entry errors but also improves the efficiency of managing multiple batteries across different stages of their lifecycle. Furthermore, a well-structured interface ensures that manufacturers can quickly access sustainability metrics and regulatory compliance data, helping them meet environmental standards without unnecessary delays.

2.4.1.2 Consumers

For consumers, the Battery Passport system serves as a tool for making informed purchasing decisions and promoting sustainable battery use. An intuitive and simple UI allows consumers to easily navigate the system, accessing critical information such as the environmental impact, performance history, and end-of-life recycling options for their batteries. By providing clear, accessible data, the UI builds trust, enhances the consumer experience, and encourages more responsible use of batteries. A well-designed interface also strengthens consumer engagement with sustainability efforts, influencing their decision-making processes toward greener choices.

2.4.1.3 Recycling Facilities

Recycling facilities rely on efficient access to data regarding battery composition, usage history, and material sourcing to ensure proper recycling processes. A user-friendly interface is crucial for recyclers, enabling them to retrieve detailed information about a battery's lifecycle and material composition effortlessly. This ensures compliance with environmental regulations and facilitates informed decisions on resource recovery, optimizing both operational costs and material efficiency. An easy-to-navigate UI also supports better workflow management, reducing administrative burdens and enhancing the overall recycling process.

A user-friendly UI is essential to the successful adoption and use of the Battery Passport system across all stakeholders. By enhancing accessibility, simplifying navigation, and promoting transparency, the system's interface can significantly improve the efficiency of data management, foster greater consumer engagement with sustainability practices, and optimize resource recovery processes in recycling.

To further support adoption, especially among first-time or nontechnical users, the system should integrate user onboarding features. Interactive tutorials, contextual tooltips, and guided workflows can help users become confident in navigating the interface without requiring external training.

2.4.2 Accessibility Considerations

To ensure the Battery Passport system is inclusive and usable for all stakeholders, including those with disabilities, accessibility principles must be integrated into the UI design. These principles not only enhance usability for individuals with specific needs but also create a more intuitive experience for all users. Below are the key considerations, organized under specific focus areas.

2.4.2.1 Sufficient Contrast

One of the foundational aspects of accessibility is ensuring sufficient contrast between text or interface elements and their background colors. High contrast improves readability for users with visual impairments, allowing them to navigate the system with ease. Interactive elements, such as buttons and forms, should feature clear, legible labels and high-contrast designs to ensure they are easily identifiable and usable by everyone [19].

2.4.2.2 Color Usage

While color is an effective tool for distinguishing information, it should never be the sole method of conveying meaning. For users who are colorblind, the addition of text-based or shape-based indicators ensures critical information remains accessible. For instance, a traffic-light-style system for alerts can be supplemented with labels such as "Error," "Warning," or "Success" to avoid reliance on color alone [20].

2.4.2.3 Interactive Elements

The functionality of interactive elements plays a significant role in accessibility. Buttons, links, and other controls should be operable with both mouse and keyboard inputs, accommodating users who cannot use a mouse due to physical limitations. Properly labeled alt text for interactive elements further ensures compatibility with screen readers, allowing visually impaired users to interact with the system effectively [21].

2.4.2.4 Consistent Navigation

Consistent navigation throughout the system is particularly important for users with cognitive impairments. A uniform layout and clear hierarchies help reduce confusion, enabling users to understand the structure of the system and move through it intuitively. For example, maintaining a consistent placement for navigation menus and employing recognizable icons can simplify the user experience [3].

2.4.2.5 Form Accessibility

Form accessibility is another critical aspect, especially for users relying on screen readers or voice commands. Input fields should be clearly labeled, with guidance provided for expected inputs. Features such as error validation messages in plain language and clearly defined instructions make it easier for all users to complete forms accurately [22].

2.4.2.6 Feedback Mechanisms

Feedback mechanisms, such as alerts or notifications, are essential for informing users when actions are completed or when errors occur. These mechanisms should be accessible through multiple modalities, including visual, auditory, and textual cues. For example, a form submission confirmation might include a text message, a sound, and a visual icon to accommodate users with different needs. Accessible feedback builds confidence and trust in the system's functionality [23].

As the Battery Passport is expected to be used in a wide range of environments including fieldwork, recycling centers, and mobile diagnostics, responsiveness is essential. Interfaces must adapt to different screen sizes and input types, ensuring seamless functionality on tablets, smartphones, and desktops alike.

Global deployment also requires the system to support multiple languages and localized terminology. Providing multilingual interfaces, region-specific regulatory references, and culturally intuitive icons ensures that the Passport is not just accessible but meaningful across international markets.

By integrating these accessibility principles under distinct, user-focused areas, the Battery Passport system can provide an inclusive and user-friendly experience for all stakeholders. Accessibility not only broadens the reach of the system but also aligns with the overarching goals of transparency and equity in the global battery supply chain.

Together, usability, accessibility, and thoughtful localization form the human layer of the Battery Passport infrastructure ensuring that its powerful data capabilities translate into real-world impact for every stakeholder in the value chain.

2.5 Operational Best Practices

The efficient operation of the Battery Passport system is contingent on adhering to a set of best practices designed to optimize energy consumption, streamline data management, and ensure uninterrupted system availability. Following these operational best practices will enable organizations to maintain a resilient, secure, and sustainable system throughout the battery lifecycle.

As the Battery Passport becomes a globally deployed system, performance and reliability must scale accordingly. These best practices not only maintain technical functionality but they also underpin user trust, regulatory compliance, and system adoption across regions and sectors.

2.5.1 Energy Management

In large-scale Battery Passport implementations, the energy consumed by sensors, data loggers, and communication devices can add up significantly. Energy-efficient solutions are necessary for minimizing power consumption without compromising performance.

Energy-efficient Sensors and Loggers: Selecting sensors and data loggers that are designed for low power consumption can significantly reduce the overall energy footprint of the Battery Passport system. Advanced sensors utilizing modern energy-saving technologies, such as MEMS (Microelectromechanical Systems), can help achieve this [3].

Low-power Communication Protocols: Utilizing low-power communication protocols, such as Bluetooth Low Energy (BLE) or optimizing cellular and Wi-Fi networks, can further reduce energy consumption. BLE is particularly useful for short-range communication, making it ideal for data transmission in localized networks with limited power resources [22].

Scheduled Data Logging and Transmission: Scheduling data logging and transmission at intervals, rather than continuously, can reduce energy usage while maintaining data accuracy. This approach prevents unnecessary power drain during low-activity periods, which is especially beneficial for systems operating in remote areas with limited access to power [24].

These strategies are especially critical for off-grid installations, remote monitoring stations, or embedded modules in electric vehicles contexts where energy budgets are tight and uptime is essential.

2.5.2 Data Handling Procedures

Data handling within the Battery Passport system requires robust practices to ensure that information remains secure, reliable, and accessible. Efficient data management not only secures system integrity but also ensures compliance with regulations.

Regular Backups: Automatic data backups are essential to prevent data loss. Regularly scheduled backups should be directed to secure cloud platforms or localized servers, with backup intervals optimized to minimize potential data gaps in case of system failures [23]

Secure Data Transmission: Encryption protocols, such as TLS, should be employed to protect data during transmission. This ensures that sensitive data such as battery composition, performance metrics, and lifecycle information remains secure from potential breaches. It is also crucial that communication channels adhere to international privacy standards, including GDPR [25].

Data Privacy Compliance: Ensuring compliance with data privacy regulations, such as the GDPR in the EU, is critical when handling personal and battery-specific data. Strict adherence to data privacy laws ensures both legal compliance and customer trust, especially when dealing with sensitive information related to battery materials and performance [22].

In an era of rising data regulation and growing consumer expectations, privacy is not a box-checking exercise; it is a cornerstone of system legitimacy. These data handling practices allow the Battery Passport to demonstrate security by design.

2.5.3 Ensuring Continuous System Availability

To maximize system uptime and prevent service disruptions, continuous system availability must be a priority. The following measures can help ensure the reliability of the Battery Passport system:

Redundancy Mechanisms: Backup servers, redundant network systems, and failover solutions are critical in preventing service interruptions during system failures or maintenance windows. Redundancy ensures that the system remains available and functional even during unexpected downtime [25].

Regular System Health Checks: Conducting regular health checks of the system can help identify and resolve potential issues before they escalate into critical failures. Health monitoring should include both hardware (e.g., sensors and loggers) and software (e.g., data handling and transmission systems) to maintain operational integrity [24].

Predictive Maintenance: Using predictive maintenance techniques, such as real-time monitoring and AI-based analytics, can detect potential issues in equipment before they cause system downtime. By monitoring the health of sensors, loggers, and network devices, organizations can replace or repair components before they fail, minimizing disruptions and optimizing operational efficiency [22].

In distributed industrial systems, downtime carries not just operational costs, but environmental and reputational ones as well. Proactive maintenance helps ensure that traceability and compliance tracking never go offline.

By adhering to these operational best practices, organizations can ensure that the Battery Passport system operates efficiently, securely, and reliably across all stages of the battery lifecycle. From optimizing energy consumption to implementing robust data handling procedures and ensuring continuous system availability, these practices collectively contribute to the long-term sustainability and operational success of the system.

Ultimately, these practices form the invisible backbone of the Battery Passport's performance, enabling it to function as a trusted, real-time intelligence system for a circular battery economy.

2.5.4 Battery Recycling

Spent lithium-ion batteries are collected from electric vehicles, electronics, and other sources and handled carefully due to both environmental concerns and safety hazards. Recyclers first sort batteries by type and chemistry, since an electric vehicle's battery pack (made of many cells, often lithium manganese oxide or other chemistries, and designed for high energy output over a long lifespan) must be managed differently than the single-cell lithium-ion batteries designed for low-power devices like smartphones and laptops. High-voltage packs are usually fully discharged before being disassembled into modules or individual cells for safe processing. These precautions help prevent short-circuits and fires, greatly reducing the risk of explosions during the recycling process.

After preliminary disassembly, the process turns to extracting materials from the battery components. The cells are typically shredded or crushed to break them apart into a mixture of metals, plastics, and black powdery electrode material. Some recycling systems include a thermal pretreatment: for example, cells may be heated in an oxygen-free furnace to about 350–600 °C to evaporate the flammable electrolyte solvent and decompose the binder (the glue holding electrode coatings). This controlled pyrolysis step neutralizes much of the battery's volatility, and the vapors (containing organics) are captured and burned off cleanly as CO_2 and water. An alternative or complementary method is to bathe the cells in an organic solvent that dissolves the electrolyte and binder; the solvent can then be distilled to recover the dissolved chemicals and reused with almost no loss. Once the batteries have been rendered inert, the remaining solids go through physical separation steps. Ferrous metal parts (like steel casings) are pulled out with magnets, and nonferrous metals are sorted using techniques such as eddy current separators and air classification. For instance, as crushed material falls through an air separator, heavy pieces like copper foil can drop out while lighter particles (the mixed oxide powder) are carried further by the air flow. In this way, aluminum and copper from the battery's current collectors are separated, leaving a fine blackish powder known as "black mass," which consists of the battery's critical mineral content (lithium, cobalt, nickel, manganese, graphite, etc.) now ready for chemical recovery (Figures 2.12 and 2.13).

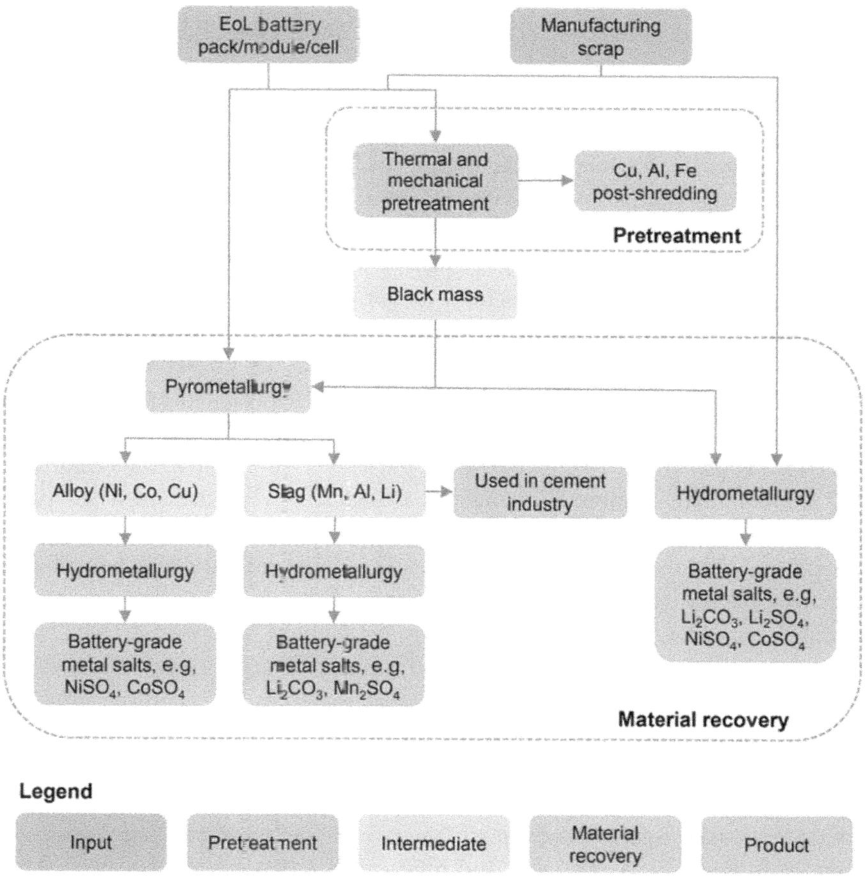

Figure 2.12 Battery recycling processes and pathways. Note: EoL, end of life; Cu, copper; Al, aluminum; Fe, iron; Ni, nickel; Co, cobalt; Mn, manganese; Li, lithium; $NiSO_4$, nickel sulfate; $CoSO_4$, cobalt sulfate; Li_2CO_3, lithium carbonate; Mn_2SO_4, manganese sulfate; Li_2SO_4, lithium sulfate. *Source:* Recycling of Critical Minerals IEA / IEA / CC BY 4.0 [28].

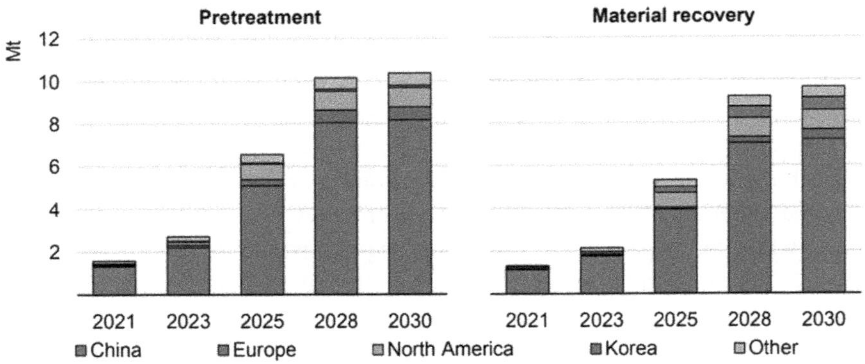

Figure 2.13 Expected battery recycling capacity based on announced projects. Note: Capacity in kilotons of cell-equivalent mass of total recyclable material. *Source:* Recycling of Critical Minerals IEA / IEA / CC BY 4.0 [28].

The next stage is material recovery from the black mass using chemical processes. Most modern recycling plants favor a hydrometallurgical approach – in simple terms, using solvents and reagents to leach metals out of the black mass and then selectively recover each one. The shredded electrode powder is dissolved in acid, which transfers valuable metals like lithium, nickel, cobalt, and manganese into a liquid solution. This is an improvement on older pyrometallurgical methods that smelt whole batteries at extremely high temperatures to produce a mixed metal alloy. Such smelting is energy-intensive and cannot capture lithium, which mostly ends up in the slag. The leachate containing the dissolved metals is then purified and treated to separate each element. Lithium can be precipitated from the solution (often as lithium carbonate) once other metals are removed. Nickel, cobalt, and manganese are typically recovered by adjusting the solution's chemistry or by solvent extraction, using special organic extractants that bind with specific metal ions. Notably, some companies have developed novel extractants that can simultaneously extract nickel, cobalt, and manganese from the leach solution in a single step, shortening the usual multistep separation process and reducing treatment cost. Through such innovations, a concentrated mixture of nickel, cobalt, and manganese can be directly obtained and converted into new battery precursor material. Meanwhile, any leftover lithium in the solution is recovered in its own stage, and the earlier-separated copper and aluminum can be sent to smelters for recycling as well. The result is that all these materials – now in the form of high-purity salts or powders – are refined to a quality suitable for making new batteries, displacing a portion of the virgin minerals that would otherwise need to be mined.

Finally comes reuse and regeneration of these recovered resources. The reclaimed lithium, cobalt, nickel, manganese, and other constituents are reintroduced into the battery supply chain, where they can be used to manufacture new electrodes and cells, essentially closing the loop of the battery's lifecycle. In addition, if some batteries or components are not fully exhausted, they may find a second life before recycling – for example, a partially spent EV battery pack might be repurposed for stationary energy storage. On the materials level, researchers and companies are also exploring "direct recycling" techniques that restore spent electrode material for reuse without fully breaking it down to basic elements. In this approach, the cathode material from a used battery is carefully extracted and rejuvenated (its crystal structure is repaired and lithium content replenished) so that it can be used again in a new battery cell. This direct regeneration is especially promising for lower-value cathode chemistries like lithium iron phosphate. In fact, recent pilot projects have demonstrated the feasibility of directly repairing and reusing Li-ion cathodes – for instance, rebuilding degraded LFP cathodes from old batteries and installing them into new cells. By combining thorough collection and safe disassembly with advanced separation, recovery, and regeneration methods, lithium-ion battery recycling can significantly reduce waste and pollution while recirculating critical materials. This process not only prevents hazardous battery components from entering the environment but also helps meet the growing demand for battery-grade metals by recovering them from spent batteries rather than relying solely on mining new resources.

2.6 Future Challenges and Innovations

The Battery Passport system is entering a critical growth phase. As demand accelerates and technologies mature, its success will depend not only on what it can do today but also on how well its designers anticipate the future. Those who create and implement this system will need to plan for sustainable scaling, data compliance, and continued interoperability in a fractured digital landscape.

2.6.1 Upgrading Equipment

Anticipating equipment upgrades in the Battery Passport system is a vital component in ensuring that the system evolves with technological advancements while remaining aligned with sustainability and circular economy goals. Successful upgrades demand careful planning to maintain compatibility with existing infrastructure, minimize operational downtime, and manage costs effectively. Additionally, as new technologies continue to emerge, sustainability becomes a key consideration in equipment upgrades, ensuring that the system remains environmentally responsible.

Several key factors should be considered when upgrading equipment within the Battery Passport framework.

2.6.2 Compatibility with Newer Technologies

One of the most critical aspects of upgrading equipment is ensuring seamless integration between new technologies and existing hardware and software systems. Upgrading equipment whether sensors, data loggers, or communication devices requires a thorough assessment of the following:

Protocols and Data Formats: New devices must support existing communication protocols and data formats to ensure interoperability across systems. For instance, the Open Charge Point Protocol (OCPP) for electric vehicles or ISO 15118 for V2G charging infrastructure is vital for ensuring data flow and compatibility in battery management systems [3].

Future-proofing Technology: Compatibility with emerging technologies, such as IoT-based devices and 5G networks, can significantly enhance system efficiency by improving data accuracy, reducing latency, and enabling real-time monitoring. A well-planned upgrade ensures the Battery Passport system can handle new technologies and data management requirements [22].

2.6.3 Cost Implications

Financial planning is another crucial factor when upgrading equipment. Although upgrading can involve significant upfront costs, the long-term benefits often justify the investment.

Initial Costs vs. Long-term Gains: The initial costs associated with upgrading hardware (e.g., advanced sensors and data loggers) should be weighed against long-term benefits like improved operational efficiency, reduced maintenance requirements, and enhanced data accuracy [23, 26]. Equipment that enhances data reliability can reduce operational disruptions and compliance issues, offering significant cost savings over time.

Cost of Operational Disruptions: Interruptions to regular operations during upgrades can be costly. By carefully scheduling upgrades during low-activity periods and using phased rollouts, organizations can minimize these disruptions and avoid data collection gaps. Temporary backup systems can also mitigate the risk of data loss during transitions.

2.6.4 Minimizing Downtime

Minimizing downtime during equipment upgrades is essential for maintaining system efficiency and data integrity. The following strategies can help:

Scheduled Upgrades During Low-activity Periods: Upgrades should be conducted during off-peak hours to limit their impact on operations. This ensures that critical processes such as compliance reporting and data transfer remain unaffected.

Phased Rollouts and Backup Systems: Phased rollouts allow for a gradual introduction of new technology, minimizing the risk of system-wide failures. Temporary redundant systems can maintain data flow during the upgrade process, safeguarding against any potential data loss [22].

To ensure a smooth and successful equipment upgrade, one could consider the following checklist:

Compatibility Checks: *Verify the compatibility of new equipment with existing systems, ensuring that hardware, communication protocols, and data formats are interoperable.*
Cost-benefit Analysis: *Conduct a detailed cost-benefit analysis to assess both the initial costs and the long-term financial gains of the upgrade.*
Scheduling Upgrades: *Plan upgrades during noncritical operational periods to minimize disruptions and ensure continuity of essential processes.*
Backup Systems: *Implement temporary redundant systems to ensure data continuity and avoid operational gaps during the upgrade.*
Sustainability Evaluation: *Assess the environmental impact and energy efficiency of new equipment, ensuring alignment with circular economy principles.*

By addressing these factors, organizations can continue to upgrade their Battery Passport systems in a way that promotes operational efficiency, sustainability, and long-term technological resilience.

2.6.5 Sustainability and Lifecycle Considerations

In the context of the circular economy, sustainability is a key driver for upgrading equipment. Organizations should prioritize energy-efficient and environmentally friendly equipment that supports long-term sustainability goals.

Energy Efficiency and Environmental Impact: Newer technologies, such as advanced sensors and IoT-enabled devices, often come with improved energy efficiency. These devices not only lower operational costs but also reduce the system's overall environmental impact, aligning with broader sustainability goals. When selecting new equipment, companies should also consider the recyclability of materials and the environmental footprint associated with their production [22].
Circular Practices in Equipment Upgrades: Adopting circular practices such as refurbishing old equipment or reusing certain components can significantly reduce costs and contribute to sustainability efforts. This approach also aligns with EU directives on promoting the circular economy in battery management [24].

2.6.6 Scalability

One of the most pressing challenges facing the Battery Passport system is scalability. As millions of batteries across global markets become integrated into the system, the volume of data generated from manufacturing and usage to recycling will expand exponentially. This surge requires infrastructure capable of handling real-time data intake and processing without degradation in performance. Scalable cloud storage, efficient APIs, and distributed blockchain systems must be adopted to ensure reliability as the system grows.

To prepare for these advancements, businesses should invest in adaptable hardware and modular system designs. These systems should be capable of handling future increases in data volume without requiring a complete overhaul of the infrastructure. Future-proofing IoT devices by ensuring they are compatible with upcoming technology standards (e.g., 5G and LoRaWAN) will also be critical for maintaining data flow and system efficiency [9].

2.6.7 Energy Efficiency

In order for the Battery Passport system to support, rather than undermine, sustainability efforts, it must be supported by energy-efficient infrastructure. Data transmission, analytics, and sensor operations all consume energy. To reduce its own footprint, the system must incorporate energy-saving designs, such as low-power IoT components, optimized logging schedules, and green cloud computing. Algorithms for data compression, deduplication, and adaptive data sampling may also reduce overhead.

Balancing energy consumption with system performance will be a central tension. As sustainability expectations tighten, energy audits of supporting digital infrastructure may become as important as lifecycle analyses of the batteries themselves.

2.6.8 Interoperability

In a global market, interoperability is key. The Battery Passport must integrate with diverse systems across manufacturers, suppliers, recyclers, and regulators. This requires alignment on data formats, standards, and exchange protocols. Without interoperability, silos will emerge, reducing transparency and undermining circularity. International standardization efforts and public–private collaboration are essential to harmonizing technical integration.

Interoperability also plays a critical role in the system's resilience. When devices, platforms, and applications communicate easily, the risk of data fragmentation or vendor lock-in is significantly reduced. It also lowers the cost of onboarding new stakeholders, supporting faster global adoption.

2.6.9 Data Security and Privacy

As the system processes and stores sensitive data, including sourcing, carbon footprints, and performance metrics, protecting that information becomes critical. Stakeholders must trust that their data is protected against breaches, misuse, or manipulation. Decentralized storage, role-based access controls, and encrypted transmission are all necessary, but maintaining privacy while enabling transparency remains a delicate balance. Regulatory frameworks like GDPR add additional layers of complexity and responsibility.

The Battery Passport system could adopt scalable, energy-efficient blockchain infrastructure capable of handling growing data demands without compromising security or performance. ***Implementing hybrid blockchain models that combine the benefits of public and private blockchains could optimize both security and transaction speed. Public blockchains ensure transparency, while private blockchains allow for faster, more controlled data processing within specific networks*** [10].

This challenge will only intensify as the Battery Passport expands to new jurisdictions with varying privacy laws. A federated approach to compliance, where core standards are paired with local adaptations, may be necessary to navigate this complexity at scale.

2.6.10 Implementation of Advanced Technologies

Incorporating advanced tools like AI, ML, and blockchain requires both infrastructure and human expertise. For many stakeholders, particularly smaller manufacturers or recyclers, resource constraints may limit access to the latest tools. Training, technical support, and standardized modular components will help democratize the use of high-tech features and reduce the risk of uneven adoption across the value chain.

2.6.11 AI and Machine Learning

AI and ML are set to redefine the Battery Passport's capabilities. These technologies will enhance real-time monitoring, enable advanced anomaly detection, and optimize data analytics throughout the battery lifecycle.

AI and machine learning models are expected to revolutionize how data is processed within the Battery Passport system...

To support the integration of AI and machine learning, infrastructure must be capable of processing complex datasets in real time... [26].

2.6.12 Enhanced Predictive Capabilities

Machine learning is one way to significantly improve lifecycle forecasting. Predictive models will anticipate degradation patterns, performance dips, or safety risks, allowing for proactive intervention. This is especially valuable for electric vehicles and grid-scale storage, where battery health directly affects reliability and cost.

To capitalize on predictive insights, manufacturers and service providers must redesign their workflows to include automated alerts, dynamic maintenance schedules, and preemptive lifecycle planning. Digital infrastructure should be capable of ingesting real-time sensor data and feeding it back into operational systems without bottlenecks. Pilot studies should focus on measuring the real-world economic and environmental gains of predictive intervention to drive adoption at scale.

2.6.13 Advanced Security Systems

AI-enabled cybersecurity will become vital for defending sensitive Battery Passport data. These systems will monitor network traffic, detect anomalies, and respond to threats in real time, offering dynamic protection far beyond static firewalls or access control lists.

As digital threat surfaces expand, security strategies must evolve beyond traditional IT models. AI-enabled systems should be continuously trained on emerging threat vectors specific to battery supply chains. In parallel, cybersecurity requirements should be embedded into Battery Passport standards, including compliance checklists, audit mechanisms, and incident response protocols tailored to decentralized data environments.

2.6.14 Real-Time Decision-Making

Emerging applications will allow the Battery Passport to act as more than a passive record. In real time, the system could adjust charge parameters, alert users to safety issues, or prompt recycling logistics based on predictive indicators. This real-time feedback loop will close the gap between monitoring and action.

For these applications to become viable, communication protocols must support low-latency feedback loops and reliable data capture even in variable network conditions. Edge devices must be equipped with autonomous decision-making logic, and battery management systems must be updated to respond dynamically to real-time inputs. These upgrades will require coordinated industry standards and rigorous testing across different deployment environments.

2.6.15 Energy-Efficient Algorithms

As part of the broader sustainability push, the Battery Passport will need to use AI models that are optimized for minimal energy usage. This ensures that the tools used to promote environmental responsibility do not inadvertently erode it.

Incorporating low-power design principles at both the hardware and software levels will be critical. Developers should prioritize model compression, selective sampling, and adaptive inference methods that reduce computational demand without sacrificing accuracy. Lifecycle assessments should include not only the energy savings of the batteries themselves but also the energy cost of the data infrastructure supporting the passport system.

2.6.16 Ethical and Transparent AI

Growing scrutiny over AI bias and privacy will require systems to demonstrate fairness, explainability, and ethical use. The Battery Passport's reliance on AI will only scale if trust is continuously earned, meaning data governance and model transparency will be critical.

To build a system users can trust, developers must embed transparency into the AI lifecycle from model design to deployment. This includes publishing training data sources, maintaining explainable logic paths, and implementing bias detection mechanisms. Regulatory oversight and independent audits will also play a vital role in ensuring AI governance keeps pace with system expansion.

2.6.17 Blockchain Evolution

Blockchain remains a foundational tool for integrity, auditability, and trust. In the future, hybrid models will become more prevalent, combining public blockchain transparency with the speed and access control of private networks.

Hybrid blockchain models should be designed with modular architecture to allow upgrades as network demands grow. Public blockchains can anchor the system with trust and transparency, while private channels enable faster, localized processing for commercial or proprietary data. Developers should also consider blockchain alternatives or supplements such as secure multiparty computation for use cases where traditional blockchain may be too energy- or resource-intensive.

The Battery Passport system should adopt scalable, energy-efficient blockchain infrastructure… [10].

2.6.18 Conclusion: Staying Adaptive

This chapter has traced the full technical architecture of the Battery Passport, from the sensors and data loggers that capture a battery's first moments in production, through the communication networks, cloud platforms, and blockchain systems that carry and protect that data, to the analytical tools and user interfaces that turn raw information into actionable insight. Yet what this architecture ultimately reveals is that no single technology is the limiting factor, the system is only as strong as the agreements, standards, and relationships that hold its many layers together.

The path forward for the Battery Passport system is defined by evolution and adaptability. It must remain agile enough to respond to challenges like scalability and security while embracing AI, blockchain, and edge computing to unlock new capabilities.

In the coming decade, the Battery Passport will be less about tracking data and more about enabling intelligence. The ability to adapt to evolving risks, regulatory landscapes, and user expectations will determine whether it becomes a permanent fixture in sustainable industry or a stalled prototype of what could have been.

In other words, in the coming decade, the Battery Passport will need to mature from a compliance instrument into an intelligence platform; one that does not merely record what happened to a battery, but actively informs what should happen next. That transition depends on convergence: of standards, of stakeholder incentives, of regulatory frameworks, and of the underlying digital infrastructure that ties them together. Perhaps most importantly, the system must earn trust iteratively. Manufacturers will not share sensitive performance data without confidence in access controls. Recyclers will not invest in integration without reliable data quality. Regulators will not rely on passport data for enforcement without auditability. Each of these relationships requires the system to demonstrate value before it can demand participation. The Battery Passport that succeeds will be one designed with awareness of what it cannot yet deliver, its trade-offs between transparency and privacy, and built with the modularity.

The Battery Passport's greatest test will be will be the willingness of competitors to share data, of regulators to harmonize requirements across borders, and of an entire industry to subordinate shortterm proprietary interests to the long-term demands of a circular economy. The infrastructure can be built. The question is whether the trust can be.

References

1 The Battery Pass Consortium (2023). Battery Passport content guidance. The Battery Pass Consortium. https://thebatterypass.eu/assets/images/content-guidance/pdf/2023_Battery_Passport_Content_Guidance.pdf (accessed 6 January 2025).

2 Monolithic Power Systems (2025). Battery management systems: major components of BMS. https://www.monolithicpower.com/en/learning/mpscholar/battery-management-systems/bms-basics/major-components-of-bms (accessed 6 May 2025).

3 Rufino Júnior, C.A., Riva Sanseverino, E., Gallo, P., and Koch, D. (2024). Towards to battery digital passport: reviewing regulations and standards for second-life batteries. *Batteries* 10 (4): 115.

4 MadgeTech (2023). Predictive maintenance in industrial settings: the role of data loggers in condition monitoring. https://www.madgetech.com/posts/blogs/predictive-maintenance-in-industrial-settings-the-role-of-data-loggers-in-condition-monitoring/ (accessed 6 July 2025).

5 Data Logger Inc (2025). Data loggers and IoT: a future of interconnected intelligence. https://dataloggerinc.com/blog/data-loggers-and-iot/ (accessed 8 June 2025).

6 MadgeTech (2025). Battery life and power management for data loggers. https://www.madgetech.com/posts/blogs/battery-life-and-power-management-for-data-loggers/ (accessed 25 May 2025).

7 SAPHI (2025). How to send sensor data to the cloud | Six smart ways. https://web.archive.org/web/20240417080410/. https://saphi.engineering/how-to-send-sensor-data-to-the-cloud/ (accessed 17 April 2024).

8 Cao, Z., Gao, W., Fu, Y., and Mi, C. (2024). Wireless battery management systems: innovations, challenges, and future perspectives. *Energies* 17: 3277. https://doi.org/10.3390/en17133277.

9 3beLiEVe (2021). Consolidated requirements for the 3beLiEVe battery pack. https://www.3believe.eu/wp-content/uploads/2021/03/D1.1-Consolidated-requirements-for-the-3beLiEVe-battery-pack.pdf (accessed 20 May 2025).

10 Guenther, L.H., Klein, V., Loef, G. et al. (2024). Second life for lithium-ion traction batteries. *Sustainability* 16 (17): 7288.

11 Circularise. Data in a digital product passport. Circularise [Internet]. https://www.circularise.com/blogs/data-in-a-dpp (accessed 28 June 2025).

12 Basan, M. (2024). How secure is cloud storage? Features, risks, & protection. eSecurity Planet. https://www.esecurityplanet.com/cloud/how-secure-is-cloud-storage/ (accessed 14 April 2025).

13 Stodt, F., Maisch, N., Ruf, P. et al. (2024). Collaborative smart production supply chains with blockchain-based digital product passports. Preprints.org. https://www.preprints.org/manuscript/202402.1194/download/final_file (accessed 13 March 2025).

14 Cai, L., Meng, J., Stroe, D.-I. et al. (2020). Multiobjective optimization of data-driven model for lithium-ion battery SOH estimation with short-term feature. *IEEE Transactions on Industrial Electronics* 67 (3): 2282–2290. https://doi.org/10.1109/TIE.2019.2902822.

15 Find Light (2025). Wireless communication protocols: NFC, bluetooth, WiFi, and more., https://www.findlight.net/blog/wireless-communication-protocols/ (accessed 14 April 2025).

16 Blümke, J. and Hof, H.J. (2023). Binding the battery to the pass: an approach to trustworthy product life cycle data by using certificates based on PUFs. *International Journal of Advanced Security* 16: 44–53. https://www.researchgate.net/profile/Julian-Bluemke/publication/372315681_Binding_the_Battery_to_the_Pass_An_Approach_to_Trustworthy_Product_Life_Cycle_Data_by_Using_Certificates_Based_on_PUFs/links/64afa151b9ed6874a515a945/Binding-the-Battery-to-the-Pass-An-Approach-to-Trustworthy-Product-Life-Cycle-Data-by-Using-Certificates-Based-on-PUFs.pdf.

17 zkPass (2023). A primer on transport layer security (TLS). Medium. https://medium.com/zkpass/a-primer-on-transport-layer-security-tls-a7495eeff004 (accessed 12 April 2025).

18 Popoola, O., Rodrigues, M., Marchang, J. et al. (2023). A critical literature review of security and privacy in smart home healthcare schemes adopting IoT & blockchain: problems, challenges and solutions. *Blockchain: Research and Applications* 5 (2): 100178. https://www.sciencedirect.com/science/article/pii/S2096720923000532.

19 Johansson, B., Despeisse, M., and Bokrantz, J. (2024). Challenges and opportunities to advance manufacturing research for sustainable battery life cycles. *Frontiers in Manufacturing Technology* 4: 1360076. https://www.frontiersin.org/articles/10.3389/fmtec.2024.1360076/full.

20 Milios, L. and Dávila, M.L. (2023). Digital product passports: implications of policy coherence and implementation towards a circular economy. https://www.researchgate.net/publication/376208532 (accessed 16 June 2025).

21 Choudhary, D., Sangwan, K.S., and Singh, A. (2024). Blockchain-enabled architecture for lead acid battery circularity. *Scientific Reports* 14 (1): 16467. https://www.nature.com/articles/s41598-024-67404-z.

22 Schulz-Mönninghoff, M. and Evans, S. (2023). Key tasks for ensuring economic viability of circular projects: learnings from a real-world project on repurposing electric vehicle batteries. *Sustainable Production and Consumption* 35: 559–575.

23 Wolf, A., Nagler, F., Daubinger, P., and Neef, C. (2024). Circular battery design: investing in sustainability and profitability. *Energy & Environmental Science* 17 (22): 8529–8544.

24 Grosa, G. (2023). Development of a data model for supporting the digital product passport in the white goods industry. https://www.politesi.polimi.it/retrieve/7c95c8a1-5a9f-42a1-b8fe-3429eabc3320/2024_7_Grosa_Tesi_01.pdf (accessed 6 April 2025).

25 Jensen, S.F., Kristensen, J.H., and Adamsen, S. (2023). Digital product passports for a circular economy: data needs for product life cycle decision-making. *Sustainable Production and Consumption* 37: 242–255.

26 Basic, F. (2023). Enabling secure and wireless battery management systems. https://tugraz.elsevierpure.com/ws/portalfiles/portal/75778689/PhD_Dissertation.pdf. (access date 22 Dec 2024).

27 Kashyap, K. (2019). Barcode, QR Code, RFID, NFC, BLE & GPS: What's better for asset tracking? *Asset Infinity* (22 November). https://www.assetinfinity.com/blog/asset-tracking-technologies.

28 IEA (2024). Recycling of critical minerals. https://iea.blob.core.windows.net/assets/3af7fda6-8fd9-46b7-bede-395f7f8f9943/RecyclingofCriticalMinerals.pdf (accessed 12 November 2025).

3

Business Models and Economic Implications

3.1 Overview of Economic Impact

(Figure 3.1)

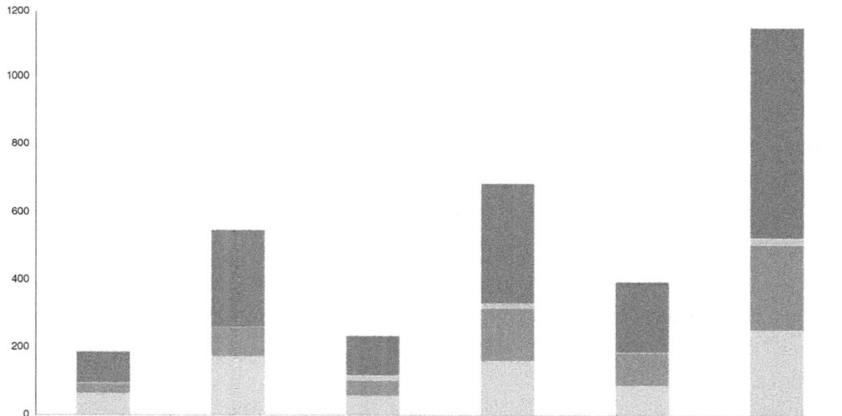

	Bp min 2037	Bp max 2037	Bp min 2040	Bp max 2040	Bp min 2045	Bp max 2045
Nickel	91	286	117	354	210	631
Manganese	0	0	13	13	0	19
Lithium	32	89	46	159	96	253
Cobalt	64	172	58	159	89	253

Figure 3.1 Recycling revenue increase. Due to the additional secondary active materials available from reducing battery leakage, we estimate that **European recyclers could increase their revenue by EUR ~ 400–1200 Mn each year** starting in 2045. *Source:* Data from Battery Pass Consortium. https://thebatterypass.eu/assets/images/value-assessment/pdf/2024_BatteryPassport_Value_Assessment_Synthesis.pdf (accessed 15 December 2026).

3.1.1 Stakeholder Roles and Relationships

The success of the Battery Passport system hinges on the active engagement of diverse stakeholders that will be economically impacted by the Battery Passport. These groups collectively drive the system's implementation, sustainability, and alignment with circular economy principles. By examining the roles and interactions of these stakeholders, we can better understand the collaborative dynamics that underpin the success of the Battery Passport system. This will help us understand who will be impacted and how, both positively and negatively, later in the chapter.

1) **Battery Manufacturers**
   Manufacturers are responsible for generating, collecting, and verifying battery lifecycle data. They ensure traceability from raw material sourcing through to battery end-of-life, maintaining transparency in line with regulatory standards and sustainability goals.
   Collaboration:
   Manufacturers collaborate with technology providers for data tracking solutions and engage recyclers to optimize battery designs for recycling efficiency.
2) **Government and Regulatory Bodies**
   Government authorities set and enforce regulations that define sustainability standards and reporting requirements. Their oversight ensures adherence to environmental and ethical practices within battery supply chains.
   Collaboration:
   Governments cooperate with international organizations (e.g., Battery Pass Consortium, Global Battery Alliance (GBA)) to harmonize regulations and work closely with industry stakeholders (manufacturers, recyclers) to align national policy goals with practical implementation measures.
3) **Consumers (Individual and Corporate)**
   Consumers drive demand for transparency and sustainability, using the Battery Passport to inform purchasing decisions. Their preferences influence the market toward environmentally responsible products.
   Collaboration:
   While consumers indirectly interact with manufacturers by shaping demand, corporate consumers (with ESG commitments) actively collaborate with manufacturers to ensure products meet sustainability criteria and reporting standards.
4) **Service Providers (Technology, Data Management, and Logistics)**
   Service providers deliver essential technological infrastructure, including data management tools, networks, Internet-of-Things (IoT) devices, and analytics platforms necessary for the Battery Passport's operation.
   Collaboration:
   These stakeholders partner with manufacturers to implement robust data-tracking infrastructure and with recyclers to supply detailed battery-composition data, improving recovery rates and facilitating regulatory compliance.
5) **Recyclers and End-of-Life Management Companies**
   Recyclers utilize battery data provided via the Passport to efficiently manage material recovery processes. Accurate lifecycle data enables them to optimize recycling methods and contribute to circular economy goals.

Table 3.1 Stakeholder groups' different roles.

Stakeholder group	Primary role	Key collaborations
Battery manufacturers	Generate, collect, and verify lifecycle data; ensure traceability and transparency.	Work with tech providers for tracking solutions; engage recyclers for recyclability.
Government and regulatory bodies	Set and enforce regulations; ensure adherence to environmental and ethical standards.	Collaborate with international orgs and industry to harmonize and implement policy.
Consumers (individual and corporate)	Use Battery Passport to inform purchases and drive market demand for transparency.	Corporate consumers work with manufacturers to meet ESG and reporting criteria.
Service providers (technology, data, logistics)	Provide technology infrastructure: data tools, blockchain, IoT, and analytics.	Partner with manufacturers and recyclers to enhance data accuracy and system integration.
Recyclers and end-of-life companies	Use passport data to optimize material recovery and recycling processes.	Coordinate with manufacturers for design improvements; partner with tech providers for data access.

6) **Collaboration:**
They closely coordinate with manufacturers to improve battery design for recycling and with service providers for accurate, timely access to lifecycle data (Table 3.1).

3.1.1.1 The Interconnected Ecosystem

The Battery Passport system thrives on the collaboration between its stakeholders. Manufacturers, government bodies, consumers, and service providers form an interdependent network where collaboration is key to achieving shared goals. Manufacturers rely on government frameworks and consumer demand to drive adoption, while service providers and governments enable the technological and regulatory infrastructure necessary for implementation.

This interconnectedness underscores the importance of fostering strong relationships across all stakeholders. The Battery Passport system is a collective effort to reshape the battery industry, aligning it with sustainability, transparency, and circular economy principles. By maintaining robust partnerships and open communication, the ecosystem can adapt to evolving challenges and opportunities, ensuring the system's long-term success (Figures 3.2–3.4).

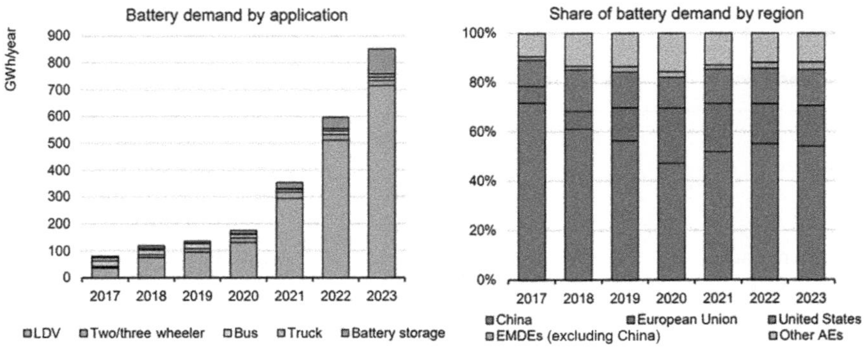

Figure 3.2 Battery demand by application and region, 2017–2023. *Source:* EV Battery Supply Chain Sustainability / IEA / CC BY 4.0.

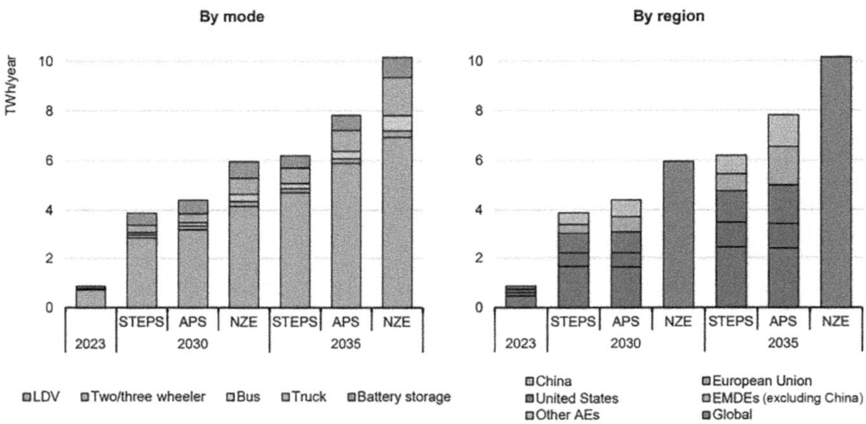

Figure 3.3 Battery demand by mode and by region in the Stated Policies Scenario, Announced Pledges Scenario, and Net Zero Emissions by 2050 Scenario, 2023–2035. *Source:* EV Battery Supply Chain Sustainability / IEA / CC BY 4.0.

3.1.2 Key Market Drivers

The factors driving Battery Passport implementation go beyond simple regulatory compliance. Economic conditions, changing consumer habits, and evolving public policy make adoption of the Battery Passport an attractive move for companies.

3.1.2.1 Shift Toward Low-Carbon Solutions

The Battery Passport plays a crucial role in the transition to low-carbon transportation, such as electric vehicles (EVs), and renewable energy storage systems. The integration of digital product passports (DPPs) aligns with international climate goals by facilitating accurate carbon footprint tracking and promoting the responsible use of materials [1].

By setting higher standards for transparency and lifecycle data, the Battery Passport system will encourage manufacturers to adopt sustainable and circular production methods. This shift may involve using recycled materials, minimizing carbon emissions

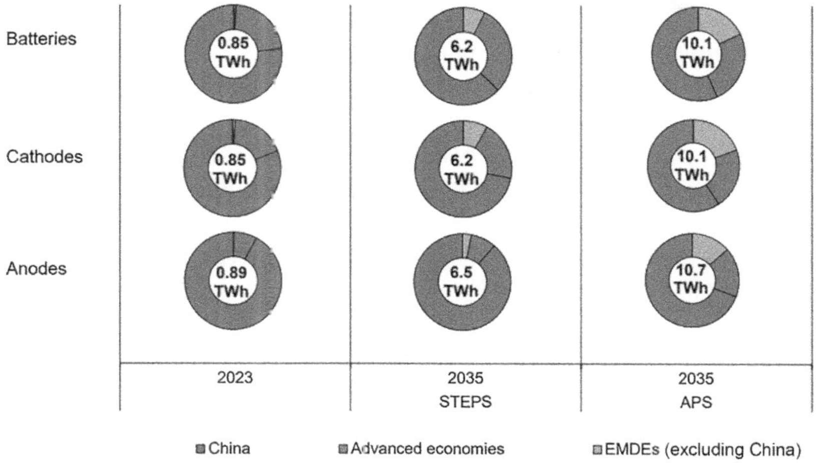

Figure 3.4 Battery and selected component production by country or region in the Stated Policies Scenario and Announced Pledges Scenario, 2023–2035. *Source:* EV Battery Supply Chain Sustainability / IEA / CC BY 4.0.

during manufacturing, and improving end-of-life management. As manufacturers aim to meet the requirements of the Battery Passport, sustainable practices will become embedded across the supply chain, from raw material sourcing to disposal [2]. *Critically, the documentation of these practices will become standard in what is currently a rather opaque industry.*

This presents a valuable opportunity for differentiation in both B2B and B2C markets. Battery manufacturers who can make verifiable claims of environmental responsibility will have the upper hand with two important customer bases: individual consumers who are concerned about sustainability and corporate buyers aiming to meet ESG goals. They also stand to gain access to capital through the growing "green finance" sector, where investors demand both sustainability and accountability.

This demand for "responsible" batteries will impact sectors such as automotive, electronics, and renewable energy, fostering a broader shift toward sustainable product development [3].

3.1.2.2 Global Push for Circular Economies

By tracking materials from production to end-of-life, the Battery Passport supports circular economy models that help businesses recover valuable resources and reduce waste. It facilitates the reuse of battery components in secondary applications, such as second-life uses for EV batteries, and enables efficient recycling processes. These factors contribute to closing the material loop and reducing dependency on virgin raw materials [4].

The Battery Passport will support circular economy principles by promoting the recovery and reuse of valuable materials from end-of-life batteries. The system will provide recyclers with essential information about the chemical composition and state of used batteries, thereby facilitating higher recovery rates of critical materials such as lithium, cobalt, and nickel. This will help reduce the reliance on virgin raw materials and lower the environmental impact of mining activities [5, 6].

What was once waste is now opportunity: Battery Passports unlock secondary value streams by making reuse, refurbishment, and remanufacturing economically viable. This data-driven circularity supports the emergence of new market actors such as second-life integrators and material recovery startups who depend on traceability to build trust and efficiency. Battery Passports therefore serve as the operational infrastructure for scaling these models across regions and industries (Figures 3.5 and 3.6).

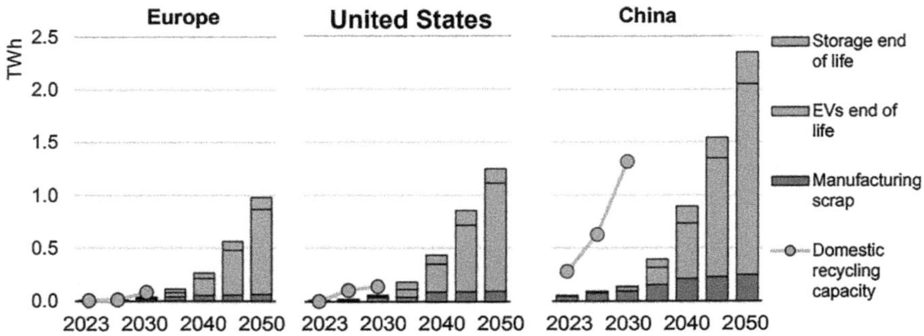

Figure 3.5 Regional available battery recycling feedstock and recycling capacity in the APS, 2023–2050. *Notes:* Available feedstock shows maximum possible feedstock volumes for recycling before any collection rates or recycling process yield losses. 85% maximum utilization rate for production capacity is assumed. Recycling capacity refers to material recovery capacity. Excludes batteries from portable electronics and e-bikes. *Source:* Recycling of Critical Minerals IEA / IEA / CC BY 4.0.

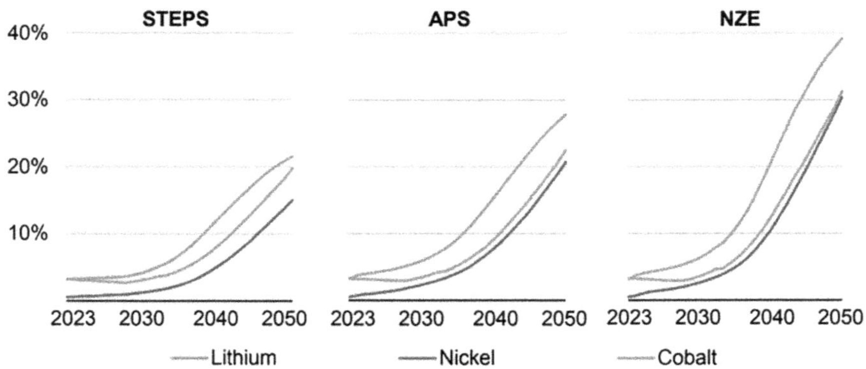

Figure 3.6 Secondary battery metals supply as a share of global demand, 2023–2050. *Notes:* Secondary supply is based on recycled metal volumes, which refer to the metal volumes recovered from recycling, accounting for collection and recycling process yield losses. Secondary supply includes EV and storage battery manufacturing scrap and end-of-life recycling. Other sources excluded. Demand includes all sources. *Source:* Recycling of Critical Minerals IEA / IEA / CC BY 4.0.

3.1.2.3 Electric Vehicle Sector

IEA projects that by 2030, there will be around 300 million EVs on the road globally. The Battery Passport will be an invaluable tool for manufacturers during this rapid expansion by facilitating compliance with environmental regulations and promoting sustainable production.

The Battery Passport supports the lifecycle management of batteries, enhancing traceability and recycling efficiency [7].

As EV adoption continues to grow, the Battery Passport will become a key component in ensuring compliance with sustainability standards. The automotive industry, especially in regions prioritizing sustainability and transparency, is expected to see increased demand for batteries with verifiable environmental credentials [8].

EVs are poised to be the largest consumer of batteries in the coming years, making the automotive sector a central force behind the adoption of Battery Passports. As the Passport becomes mandatory in Europe for EVs, it will rapidly become integral to original equipment manufacturers' (OEMs) digital supply chain strategies. It enables real-time data flows that support regulatory compliance, performance monitoring, and aftersales optimization. Beyond emissions targets, Battery Passports allow automakers to differentiate on verified circularity, resale value, and long-term service guarantees.

As EVs move into the mainstream, the Battery Passport evolves from a regulatory requirement into a strategic tool that supports product stewardship and builds customer trust. Automakers are no longer judged solely by tailpipe emissions but by the sustainability and transparency of their entire battery supply chain. With the global EV market expanding rapidly, this demand accelerates the scale and impact of the Battery Passport, positioning it as a critical enabler of the industry's sustainable transformation (Figure 3.7).

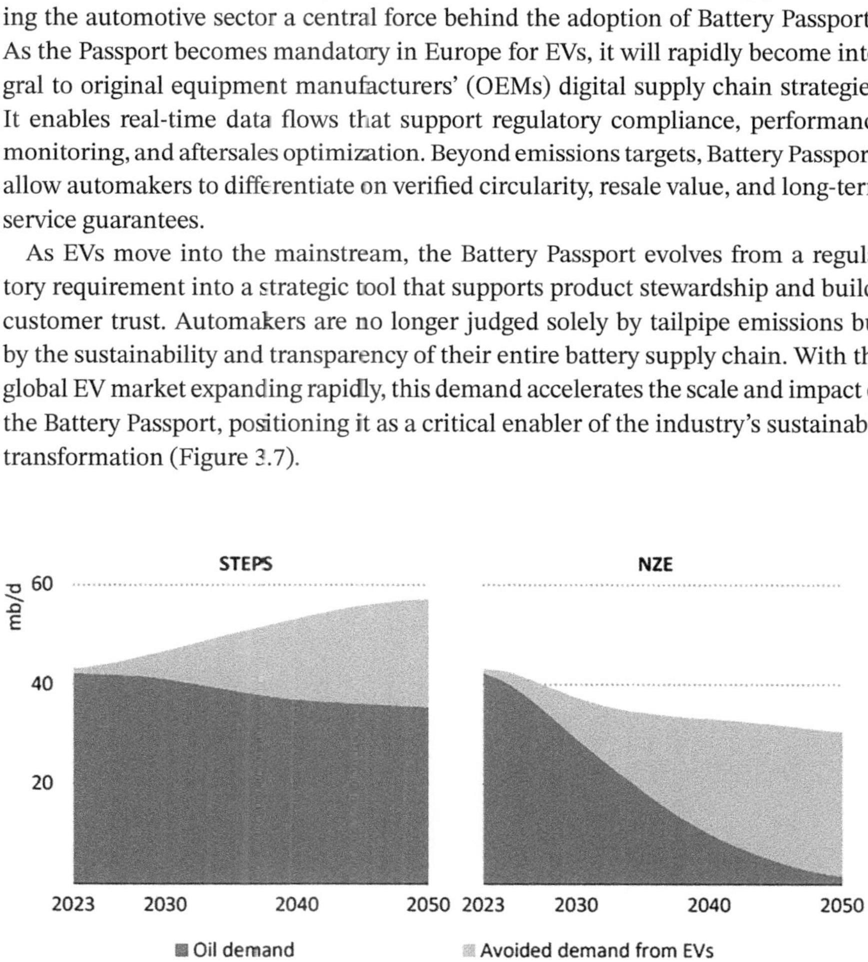

Figure 3.7 Oil demand in road transport and savings from EVs by scenario, 2023–2050. *Battery use in EVs significantly curbs future oil demand growth from road transport in the STEPS and avoids nearly 30 mb/d (million barrels per day) of oil use by 2050 in the NZE Scenario.* *Source:* Batteries and Secure Energy Transitions (IEA) / IEA / CC BY 4.0.

3.1.2.4 Lithium-Ion Battery Market

The lithium-ion (Li-ion)battery market is projected to grow by over 30% annually from 2022 to 2030, driven by increasing demand for EVs and renewable energy storage solutions. By 2030, the market could reach 4.7 TWh, with an estimated value exceeding $400 billion. The Battery Passport is expected to play a pivotal role in ensuring the sustainability of this growth by providing detailed tracking of battery components and promoting compliance with environmental regulations [9].

Batteries used in renewable energy storage systems are increasingly important for balancing supply and demand in energy grids. The Battery Passport provides the necessary information to align these systems with global sustainability goals, promoting lifecycle management and improving resource recovery from end-of-life batteries [10].

The Battery Passport will provide detailed information about the sustainability and lifecycle of batteries, including material sourcing, carbon footprint, and recycling potential. This transparency is expected to increase demand for batteries that meet high environmental and social standards. The growing adoption of EVs and energy storage solutions will further drive this trend, as industries align with global sustainability targets, such as the Paris Agreement. The Battery Passport will act as a certification tool that signals compliance with sustainability goals, thus influencing purchasing decisions [11] (Figure 3.8).

Although the BP will not be mandatory in all Li-ion utilizations, the explosive growth of the Li-ion battery market still brings with it increasing investor scrutiny around ESG risks, including sourcing ethics, carbon intensity, and end-of-life impacts. Battery Passports provide the infrastructure for managing these risks at scale, offering traceability tools that institutional investors, regulators, and global buyers increasingly demand. As growth accelerates, Passport adoption will also be a prerequisite for securing capital, entering regulated markets, and competing on sustainability, not just capacity. Thus, the growth of demand for batteries in all

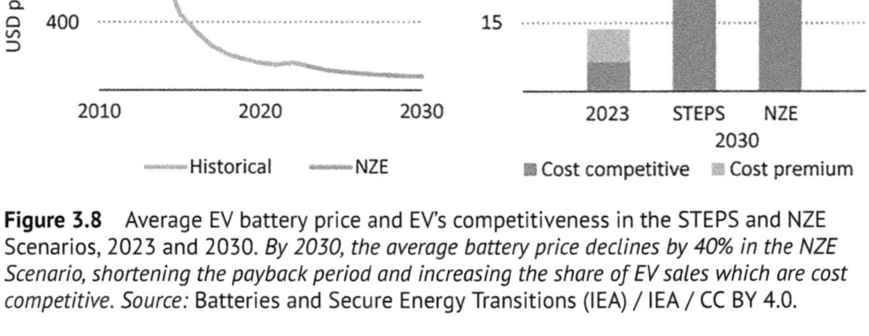

Figure 3.8 Average EV battery price and EV's competitiveness in the STEPS and NZE Scenarios, 2023 and 2030. *By 2030, the average battery price declines by 40% in the NZE Scenario, shortening the payback period and increasing the share of EV sales which are cost competitive. Source:* Batteries and Secure Energy Transitions (IEA) / IEA / CC BY 4.0.

kinds of fields, other than simply mobility, will also create a growth in the usage of Battery Passports, which is a key growth factor for the initiative.

3.1.2.5 Competitive Advantage for Early Adopters

Implementation of the Battery Passport will be accelerated by its potential as a differentiator, since early adopters and innovators stand to reap long-term benefits.

Companies that implement the Battery Passport and demonstrate a commitment to sustainability can gain a competitive edge. Early adopters can leverage the data provided by the system to enhance their brand image, optimize product offerings, and meet the increasing demands of sustainability-conscious consumers and regulators. This will enable them to differentiate themselves in the market and potentially command premium prices for certified sustainable products [12].

We can see early evidence for this in the case of Volvo, who have already implemented their own early version of the Battery Passport. While it does not include all the features we can expect from later iterations, it is a powerful way for the company to signal their commitment to sustainability and innovation. Chinese manufacturers may also roll out a robust Battery Passport system ahead of their European competitors to secure their current advantage.

The Battery Passport creates asymmetry in the market: companies that adopt early levels of transparency beyond legally required minimums will avoid future complications, gain reputational benefits, and access sustainability-focused contracts ahead of their competitors. These first movers are better positioned to shape industry standards, influence policy, and build scalable digital compliance systems. In contrast, late adopters will face higher adjustment costs, loss of market access, and declining trust.

3.1.2.6 Opportunities for New Market Entrants

As the demand for battery data and sustainability reporting grows, service providers specializing in IT solutions, data analytics, and sustainability consulting will face increasing competition. Companies offering cost-effective and innovative solutions for meeting Battery Passport requirements will likely see a surge in business, prompting ongoing technological advancements in the sector [13].

The Battery Passport will require an entire ecosystem of related services. The digital and physical infrastructure supporting the system needs to be designed, maintained, and eventually upgraded, leading to new opportunities in the IT industry. An unprecedented abundance and richness of data will be a boon to companies offering services such as sustainability consulting, data analytics, training, and battery lifecycle management. These firms can help manufacturers meet compliance requirements and optimize sustainability practices, creating a market for specialized services that support the Battery Passport ecosystem [11].

While large firms will be able to leverage their experience in consulting, supply chain, and software solutions, we also expect that these emerging needs will open the door for startups and boutique firms to deliver niche solutions with more flexibility. Demand for third-party platforms, modular tools, and consulting services will fuel innovation and competition.

Table 3.2 Market drivers and strategic implications of Battery Passport adoption across the battery value chain.

Market driver	Stakeholders impacted	Strategic implications
Low-carbon accountability	*OEMs, battery manufacturers, investors, regulators*	*Carbon transparency becomes a differentiator; access to green finance and low-carbon markets*
Circular economy integration	*Recyclers, second-life operators, material recovery startups*	*Enables new business models based on reuse, refurbishment, and recycling*
Growth in the lithium-ion battery market	*Cell manufacturers, component suppliers, investors*	*Sustainability becomes a prerequisite for scaling production and securing capital*
EV market expansion	*Automotive OEMs, fleet operators, governments*	*Battery traceability becomes essential for compliance, resale value, and consumer trust*
Competitive advantage for early adopters	*All value chain players, especially Tier 1 suppliers and OEMs*	*Early movers can access premium markets and establish leadership in sustainability*
Demand for data and compliance services	*Tech firms, consultants, software vendors, compliance auditors*	*Expands the market for digital tools supporting traceability, auditing, and lifecycle analytics*

A wide range of economic players, from massive firms to lean start-ups, stand to gain from this economic opportunity, further incentivizing adoption of the Battery Passport (Table 3.2).

3.1.3 Shifts in Consumer Behavior, the Most Important Market Driver

As consumers increasingly prioritize sustainability, the Battery Passport's capacity to help companies address their concerns is a major market driver. Growing awareness of environmental issues is reshaping consumer behavior across markets, driving up demand for batteries through the adoption of products such as EVs. Moreover, the same eco-conscious consumers creating this demand are increasingly aware of the environmental and ethical issues associated with battery manufacturing. Manufacturers are therefore under double pressure: to remain competitive, they must rapidly scale up production while simultaneously improving their sustainability and transparency practices. Companies that fail to meet evolving environmental standards will falter, while those on the cutting edge will gain access to the most discerning consumers.

The Battery Passport is a key tool that will help manufacturers rise to the challenge. It is designed to promote transparency and collect information about the environmental footprint of batteries throughout their life cycle, providing key data that will help companies improve their practices on a large scale and demonstrate accountability to consumers.

This section explores how changing consumer preferences are influencing demand for Passport-certified products, as well as the strategies companies can adopt to engage customers effectively.

The shift in consumer preferences toward sustainable products has been well-documented. A report on sustainability and circularity metrics for EVs discusses how the growing emphasis on sustainable products is influencing battery markets, as consumers increasingly demand transparency about the environmental impact of their purchases [14].

3.1.3.1 Willingness to Pay for Certified Products

Many consumers are willing to pay a premium for products certified as environmentally friendly or ethically sourced. Brands such as Patagonia capitalize on their reputation for sustainability to capture market share in premium segments and build brand loyalty – and this strategy could be adopted by battery manufacturers as well. A study found that a significant portion of consumers expressed a readiness to pay more for battery-powered products, such as EVs and electronics, if they were certified under a Battery Passport system. Particularly in the context of the EV industry, where the average buyer is already highly motivated by environmental concerns, the Battery Passport presents companies with an opportunity to build trust and loyalty [15].

Battery Passport compliance could unlock value-added positioning, allowing businesses to optimize their profit margins through trust, quality, and transparency. On the other hand, companies that fail to build a reputation for responsibility may be excluded from retail platforms, financial products, or corporate procurement lists that prioritize ESG performance. The latter is particularly important, as corporate car fleets make up a significant proportion of car purchases globally.

3.1.3.2 Perception of Environmental Benefits

Shifting consumer behavior is one of the most influential market drivers for the Battery Passport. As sustainability becomes a core value for buyers, perception carries real economic weight. The transparency provided by the passport helps consumers make informed decisions, aligning with the growing trend of eco-conscious consumerism. The digital documentation of battery lifecycles, including material sourcing and recycling efforts, allows consumers to support products that prioritize sustainability [16].

This verified transparency gives brands a competitive edge, turning the Battery Passport into a powerful marketing tool and a source of brand equity. In high-profile sectors like EVs, where lifecycle impact is under both public and regulatory scrutiny, this matters more than ever. Companies that embrace clear, credible environmental disclosures stand to gain trust and loyalty, while those that do not risk reputational damage, consumer skepticism, and long-term erosion of market position.

Finally, it is important to note that despite the growing demand for more sustainable products, this trend has recently encountered significant headwinds. Inflation, driven by factors such as supply chain disruptions, rising energy costs, and geopolitical instability, has slowed consumer spending, especially in sectors where sustainable options often come with a higher price tag. As a result, customers are more likely than before to prioritize affordability over environmental impact, leading to a temporary deceleration in the adoption of eco-friendly goods and services.

However, this slowdown does not signal a reversal of the broader shift toward sustainability. On the contrary, structural changes in regulation, investment, and consumer awareness maintain the long-term importance of sustainable development. Once economic conditions stabilize, we expect that the momentum will return, supported by growing pressure on companies to disclose environmental performance, the rise of green finance, and increasingly stringent policy frameworks worldwide.

3.1.3.3 Engaging Consumers in Battery Passport Awareness

Effective consumer engagement strategies can significantly influence the adoption of Battery Passport-certified products. Reports indicate that clear communication and educational initiatives are crucial for increasing consumer awareness and understanding of Battery Passports [17]. Companies can use the Battery Passport to create brand awareness, making the passport a powerful tool to engage with customers and sell more products.

3.1.3.4 Education and Communication Strategies

To engage consumers effectively, stakeholders could use digital platforms, social media campaigns, and public outreach initiatives to educate the public on how Battery Passports contribute to sustainability goals. Such strategies help raise awareness about the benefits of transparent supply chains and circular economy practices, fostering a demand for certified products.

Well-executed awareness campaigns can accelerate adoption and expand the market for Battery Passport-certified products. This proactive outreach helps convert general interest in sustainability into purchasing behavior. On the flip side, poor communication or technical messaging may result in confusion or apathy, dampening the commercial upside of certification and stalling return on sustainability investments.

3.1.3.5 Incentives for Adoption

Financial incentives such as tax rebates, subsidies, or discounts for verifiably sustainable batteries can further encourage adoption and good behavior from the manufacturers. Incentives for purchasing certified EVs can help mitigate the higher upfront costs associated with these products, making them more attractive to consumers by reducing the price barrier [18].

These tools help bridge the affordability gap, making Battery Passport-certified goods more competitive in mainstream markets. For governments and companies alike, offering targeted incentives generates broader uptake, faster returns on investments (ROI) on infrastructure investments, and a more robust circular economy. Without such measures, early adoption may remain limited to high-income consumers, slowing systemicchange.

3.1 Broader Market Implications for Digital Product Passports (DPP)

1) ***Customer Demand for Verified Sustainability***
Sustainability is no longer a niche concern. Consumers and corporate buyers are demanding credible, granular data on the environmental and social impact of products. DPPs enable this by embedding traceability and impact metrics directly into products, supporting claims of social and environmental responsibility with real data. This is especially potent in high-visibility sectors like fashion, electronics, and automotive, where purchasing decisions are increasingly values-driven.
2) ***Regulatory Momentum as a Market Shaper***
While regulation is not the only driver, it is a powerful accelerant. The European Union's ***Sustainable Products Initiative (SPI)*** *and* ***Ecodesign for Sustainable Products Regulation (ESPR)*** *are creating a clear policy trajectory toward mandatory DPPs for high-impact products. These rules have the power to reshape markets by setting clear standards for product design, data transparency, and circularity. Companies that adapt early will win access to tightly regulated markets and avoid costly retrofits later.*
3) ***Circular Economy as a Business Imperative***
DPPs are also gaining traction as tools for enabling circular business models such as repair, reuse, refurbishment, and material recovery. By making product composition, repairability, and recycling pathways visible and verifiable, DPPs lower transaction costs and unlock new revenue streams. Companies see growing financial logic in designing for longevity and traceability, especially as virgin material costs rise and carbon pricing becomes a factor.
4) ***Technology Enablement and Data Infrastructure Growth***
Advances in IoT, AI, and blockchain are removing friction and reducing the cost of DPP implementation. Smart tagging, real-time monitoring, and decentralized data storage now make it feasible to track products across their full lifecycle. This opens new markets for tech providers while reducing barriers for industries looking to comply. The digital infrastructure emerging around Battery Passports is laying the groundwork for broader DPP adoption across sectors.
5) ***Investor and Supply Chain Pressure***
Institutional investors are increasingly scrutinizing ESG metrics – not just from final brands, but across the full supply chain. DPPs provide verifiable data that suppliers can use to demonstrate compliance and resilience. This drives adoption from the bottom up, as Tier 2 and Tier 3 suppliers seek to maintain relationships with downstream partners who face disclosure obligations. In parallel, green finance incentives are emerging for products with traceable, low-impact lifecycles.

3.1.4 Evolving Infrastructure Needs: A Business Opportunity

As the Battery Passport system develops and scales, its underlying infrastructure must evolve to support the industry. This creates business opportunities for numerous players.

3.1.4.1 Advancements in IoT and Sensor Technology

With the growth of IoT devices and advanced sensors, the infrastructure must support increased data collection and real-time processing. Future systems will need to handle high volumes of data generated by sensors monitoring battery performance, material degradation, and environmental conditions. Enhanced bandwidth and storage capabilities will be necessary to process this information efficiently and ensure timely data accessibility.

Business Opportunity: *Companies specializing in miniaturized sensors, real-time monitoring devices, or environmental diagnostics stand to gain early traction. Meanwhile, cloud and edge computing providers benefit from the need to process vast sensor data efficiently and close to the source. System integrators and platform vendors who can offer end-to-end IoT solutions – hardware, connectivity, analytics, and compliance formatting – will be in high demand. In short, tech firms that help translate physical battery behavior into verifiable digital records will be key winners as data becomes central to battery value and compliance.*

Preparation Strategy: Invest in scalable and flexible infrastructure that can accommodate expanding data requirements without compromising on performance. This may include using edge computing to process data closer to the source, thus reducing latency and optimizing network bandwidth [19]. R&D in sensor technology may also be a worthwhile investment (Figure 3.9).

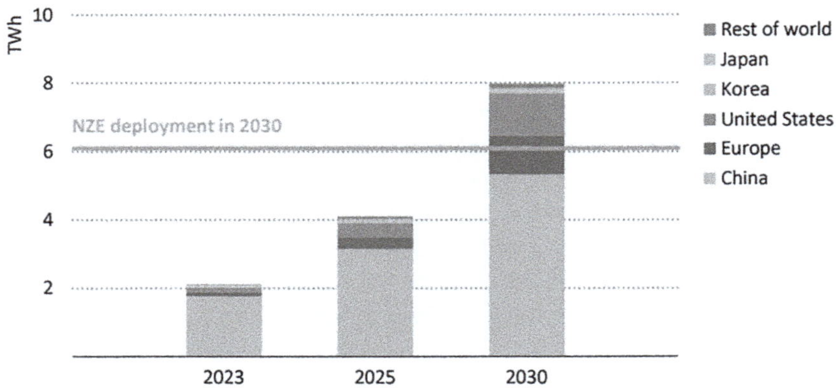

Figure 3.9 Maximum lithium-ion battery manufacturing capacity by region and needs in the NZE Scenario, 2023, 2030, and 2050. *If announcements materialize, battery manufacturing capacity more than triples by 2030 and is sufficient to meet future battery demand in the NZE Scenario. Source:* Batteries and Secure Energy Transitions (IEA) / IEA / CC BY 4.0.

3.1.4.2 Integration of AI and Machine Learning

AI and machine learning will increasingly be used for battery performance optimization, predictive maintenance, and data analytics. It is important to note that most battery manufacturers already engage in such practices on their batteries, but the passport could take this to the next level.

***Business Opportunity*:** The Battery Passport creates strong demand for AI solution providers who can deliver performance forecasting, predictive maintenance, and anomaly detection at scale once high connectivity is rolled out. AI startups, data science firms, and cloud service providers stand to gain by building machine learning models tailored to battery lifecycle optimization. As real-time analytics becomes a compliance and operational necessity, vendors offering pretrained models, analytics platforms, or ML-as-a-service solutions will be well positioned to win contracts from OEMs, recyclers, and regulatory agencies.

Preparation Strategy: Integrate AI-friendly hardware and software, such as graphics processing units (GPUs) and specialized cloud-based services, capable of supporting deep learning models and large datasets. Consider using federated learning approaches to train models across distributed data sources while preserving data privacy [20].

3.1.4.3 Blockchain for Enhanced Security and Transparency

Blockchain technology will continue to be integral to the Battery Passport system by ensuring secure and transparent data sharing across the supply chain. The evolving nature of blockchain will require infrastructure capable of supporting decentralized data management, with higher transaction speeds and lower energy consumption. The ability to integrate both public and private blockchains will be key to meeting varying regulatory and data privacy requirements across regions.

It is important to note that blockchain developers are already taking the lead in developing Battery Passport infrastructure. Much of the existing literature on the subject comes from these developers, indicating that they are proactively investing labor and resources in this space. Companies like *Minespider, dltledgers*, and *Circularise* are already deploying blockchain-based Battery Passports that offer traceability, lifecycle transparency, and secure compliance tools for global supply chains.

***Business Opportunity*:** *Blockchain developers, protocol designers, and decentralized identity providers have a prime opportunity to shape the security backbone of the Battery Passport. Businesses that offer scalable, energy-efficient blockchain solutions will be in demand to support traceability, authenticity, and tamper-proof compliance records. Especially valuable are firms that can deliver permissioned blockchain frameworks for regulatory use or cross-chain interoperability tools for multinational supply chains.*

Preparation Strategy: Invest in energy-efficient blockchain protocols, such as Proof-of-Stake or hybrid consensus mechanisms, that reduce the environmental impact while providing scalability. Additionally, prioritize interoperability to enable seamless integration with existing systems and different blockchain networks [21].

3.1.4.4 Big Data Analytics for Lifecycle Management

As the volume of data collected increases, there will be a growing need for advanced analytics to derive actionable insights. If such data is shared with the passport, Big Data techniques can be used to analyze battery usage patterns, recycling trends, and material recovery rates, supporting the optimization of the entire lifecycle from production to disposal.

Business Opportunity: *Massive volumes of passport data create a lucrative space for big data analytics companies, lifecycle management software firms, and advanced analytics consultants. These players can monetize tools that help manufacturers, regulators, and recyclers make sense of battery behavior across space and time. Real-time dashboards, risk scoring engines, and optimization algorithms will be essential tools and represent scalable, high-margin service offerings for data-driven enterprises.*

Preparation Strategy: Incorporate data lakes and distributed storage solutions that can handle diverse data formats and large volumes. Implement analytics platforms capable of real-time processing and predictive modeling to identify optimization opportunities throughout the battery lifecycle [6].

3.1.4.5 Edge Computing and 5G Integration

The use of edge computing will allow for data processing closer to where the data is generated, such as on IoT sensors attached to batteries. Eventually, if batteries end up being highly connected, the deployment of 5G networks will enable reliable data transmission and faster response times for real-time monitoring applications.

Business Opportunity**:** *As batteries increasingly communicate* via *real-time telemetry, edge computing firms, 5G network operators, and device makers have the chance to lead the infrastructure buildout. The key winners will be those who can offer ruggedized, low-latency edge hardware or bundle connectivity and compute power into seamless edge-cloud solutions for battery OEMs and fleet managers needing instant compliance and performance insights.*

Preparation Strategy: Develop hybrid computing architectures that utilize both cloud and edge computing to optimize processing and storage. Leverage 5G connectivity to support high-speed data transfer and improve the reliability of real-time applications [22].

3.1.4.6 Data Center Requirements and Expansion

As the Battery Passport system scales, the demand for robust data centers capable of managing vast amounts of information will grow. These centers will need to process, store, and analyze real-time data from millions of batteries worldwide, creating a need for high-performance infrastructure. Establishing regional data centers can reduce latency, ensure compliance with local data regulations, and provide stakeholders with timely access to insights. To align with sustainability goals, these data centers should prioritize energy efficiency through the use of renewable energy sources and innovative cooling technologies. Green data centers, designed with carbon-neutral operations in mind, will play a critical role in minimizing the system's environmental footprint while meeting its computational needs.

Business Opportunity: *The Battery Passport's global scale demands new generations of green data centers, presenting a massive opportunity for sustainable infrastructure providers, cloud operators, and modular data center builders. Enterprises that can deliver low-carbon, high-performance computing environments, especially in strategic regions like Africa, Southeast Asia, and Latin America, will secure long-term contracts to host compliance-critical data with minimal environmental footprint.*

Preparation Strategy: Invest in modular, energy-efficient data centers powered by renewable energy. Utilize advanced cooling systems, such as liquid cooling, and implement regional facilities to reduce latency and improve compliance with local regulations.

3.1.4.7 Technology Implementation Facilities

The implementation of the Battery Passport system will require dedicated facilities for the production and integration of advanced technologies. These facilities must be equipped to manufacture high-precision components and ensure their seamless integration into battery systems. If regulatory and market forces favor widespread retrofitting of older batteries, a need for specialized facilities will emerge. Standardization and modularity in equipment design will be essential to streamline manufacturing processes and ensure interoperability across diverse battery models and applications.

Business Opportunity: *The need to integrate sensors and smart modules into batteries creates demand for specialized tech manufacturing hubs. Contract manufacturers, original design manufacturers (ODMs), can capitalize by setting up agile, modular production lines that cater to Passport hardware needs. There is also a niche for firms focused solely on adapting legacy battery systems to meet Passport requirements, especially in industrial and commercial energy storage.*

Preparation Strategy: Focus on establishing modular production facilities capable of scaling efficiently. Implement quality assurance processes and standardization protocols to ensure seamless integration across various battery types.

3.1.4.8 Logistics and Supply Chain Infrastructure

Effective logistics and supply chain infrastructure will be key to the success of the Battery Passport system. Tracking batteries throughout their lifecycle requires advanced transportation monitoring systems and safety mechanisms integrated with real-time data capabilities. Regional lifecycle management hubs equipped with recycling facilities and data integration technologies can optimize resource recovery and waste reduction. Spent batteries are considered hazardous waste, and as such, their handling carries special logistical concerns that present an opportunity for specialized companies. By leveraging smart logistics solutions, the system can minimize transportation inefficiencies and reduce its overall environmental impact and safety risks.

Business Opportunity: *Smart logistics companies, fleet management software providers, and supply chain visibility platforms are uniquely positioned to support*

Passport-enabled tracking of batteries from factory to end-of-life. Players that integrate real-time GPS, condition monitoring, and automated compliance reporting and safety mechanisms will win clients looking to streamline operations and demonstrate circularity. There is also growth potential in lifecycle logistics services, focused on routing batteries for second life or recycling based on Passport data.

Preparation Strategy: Establish regional hubs with integrated logistics and recycling capabilities. Deploy IoT-enabled transportation tracking systems to optimize routes, improve efficiency, and enhance resource recovery. Most importantly, have fleets and logistics ready to transport large amounts of dangerous and flammable spent batteries.

3.1.4.9 Workforce Development and Training Centers

The adoption of cutting-edge technologies and advanced end-of-life logistics within the Battery Passport system necessitates a highly skilled workforce. Establishing workforce development and training centers will be crucial to equip personnel with the expertise required to manage and operate the system effectively. These programs should focus on building technical proficiency in various areas such as data analysis, security, and lifecycle management. Additionally, training on sustainability practices and regulatory compliance will ensure that the workforce can align operational goals with environmental and industry standards.

***Business Opportunity*:** *With new regulatory and technical demands, there is strong demand for vocational schools, EdTech platforms, and corporate training providers offering certified programs in battery traceability, data systems, and compliance tech. Companies that can deliver industry-recognized certifications, especially in emerging markets, stand to benefit from both government support and private-sector upskilling mandates tied to Battery Passport adoption.*

Preparation Strategy: Develop specialized training programs in schools or companies and certifications tailored to the technological and regulatory needs of the Battery Passport system. Collaborate with academic institutions and industry leaders to ensure workforce readiness.

3.1.4.10 Regulatory and Compliance Frameworks

Supporting compliance with global and regional regulations requires a robust infrastructure of auditing tools and data validation systems. These tools should integrate seamlessly into the Battery Passport system to automate compliance reporting and ensure adherence to evolving standards. Collaborative platforms that bring together government agencies and industry stakeholders can help standardize processes and foster alignment across regions. Such frameworks will not only ensure regulatory compliance but also build trust among stakeholders by demonstrating a commitment to transparency and accountability.

***Business Opportunity*:** *There is a clear niche for regulatory tech firms, compliance automation startups, and standard-setting consultancies and certification bodies that can create tools for monitoring, reporting, and verifying Battery Passport data and compliance. Companies offering plug-and-play audit platforms or cross-border compliance modules will become essential partners for manufacturers and exporters seeking to navigate evolving regulations without building custom systems.*

Preparation Strategy: Implement automated compliance monitoring systems integrated with real-time reporting tools. Collaborate with regulatory bodies to create standardized frameworks that adapt to regional and global requirements.

3.1.4.11 Pilot Programs and Test Facilities

Before full-scale deployment, pilot programs and test facilities will be essential for evaluating new technologies and infrastructure components. These facilities provide a controlled environment to assess the performance, interoperability, and scalability of systems such as blockchain networks and IoT devices. By simulating real-world conditions, pilot programs can identify potential bottlenecks and challenges, allowing for adjustments before widespread implementation. Test facilities also foster innovation by enabling stakeholders to experiment with novel approaches to data management and battery lifecycle optimization.

Business Opportunity: *Innovation labs, engineering consultancies, and testing-as-a-service providers will find new revenue streams in developing and operating testbeds for Passport technologies. These facilities can serve OEMs, governments, or certification bodies looking to validate hardware, software, or entire compliance workflows before commercial rollout. Stakeholders who offer real-world simulation capabilities and multistakeholder sandbox environments will hold strategic value.*

Preparation Strategy: Establish regional test centers equipped with simulation tools to replicate real-world conditions. Use pilot programs to gather feedback from stakeholders and refine processes before full deployment.

3.1.4.12 Integration with Renewable Energy Systems

The Battery Passport system must integrate seamlessly with renewable energy systems to optimize lifecycle management and align with sustainability objectives. This integration requires infrastructure capable of linking battery data with smart grids and energy storage hubs, ensuring efficient energy use and recovery. Technologies such as advanced energy management software and IoT-enabled communication systems will be critical for real-time coordination between the Battery Passport system and renewable energy grids. These advancements will enhance the alignment between battery lifecycle data and global renewable energy goals, promoting a more sustainable energy ecosystem. Such advanced connectivity could help understand the real carbon footprint of a battery or vehicle over its lifecycle, accounting for the carbon footprint of its energy use. However, such an advancement is not expected anytime soon.

Business Opportunity: *The Battery Passport's ability to tie into smart grids creates space for* ***energy platform developers, virtual power plant operators,*** *and* ***distributed energy resource (DER) aggregators.*** *Businesses that connect Passport data with grid performance, energy trading, or storage optimization platforms can unlock premium services like* ***real-time carbon tracking, green energy incentives,*** *or* ***battery-grid arbitrage.*** *This blurs the line between clean tech and fintech and opens powerful new markets.*

Preparation Strategy: Invest in IoT-enabled energy management systems and ensure interoperability with smart grids. Develop predictive analytics tools to optimize energy usage and recovery within renewable systems.

3.1.4.13 Cybersecurity Infrastructure

The secure handling of sensitive battery data necessitates a robust cybersecurity infrastructure. This includes the establishment of dedicated security operations centers equipped with tools for real-time threat detection and mitigation. Blockchain technology will play a pivotal role in ensuring data integrity, but additional safeguards, such as end-to-end encryption and multifactor authentication, are necessary to protect decentralized networks. As the system scales, proactive measures must be taken to address emerging cybersecurity risks, ensuring that the infrastructure remains resilient and capable of protecting stakeholder data from potential breaches or malicious attacks.

Business Opportunity: *The need to protect distributed, high-value battery data makes this a high-growth area for* ***cybersecurity vendors, blockchain security firms,*** *and* ***industrial threat intelligence platforms.*** *Providers that specialize in securing IoT networks, preventing data tampering, or building real-time incident response systems will be indispensable to Passport operators. Those offering* ***zero-trust architectures, blockchain forensics,*** *or* ***data residency compliance*** *can differentiate themselves in a highly regulated field.*

Preparation Strategy: Implement a multilayered security framework incorporating advanced encryption, blockchain-based auditing, and real-time threat detection systems. Regularly update and test cybersecurity protocols to stay ahead of potential threats.

A comprehensive study by the Battery Pass Consortium, backed by the German Federal Ministry for Economic Affairs and Climate Action (BMWK), offers the first detailed evaluation of the advantages and challenges of implementing digital Battery Passports as required by the EU Battery Regulation. The study highlights how Battery Passports can improve transparency, sustainability, and efficiency across the battery value chain, while also pinpointing significant obstacles and suggesting recommendations for successful implementation.

Key Quantitative Findings

Area	Details
Cost reductions	Procurement and technical testing costs for independent operators could drop by 2–10%. Preprocessing and treatment costs in recycling could decrease by 10–20% due to reduced sampling needs.
Environmental impact	Extending battery lifespans through passports could save 370,000 to 1.3 million tons of CO_2 annually across the EU. Recycling rates might improve by 1–2%.
Strategic resource recovery	Incorporating Battery Passports into vehicle deregistration and export systems could recover an estimated 5–20% of the active material demand for EV batteries in Europe by 2045.

Strategic and Operational Implications

Area	Details
Business value	Battery Passports facilitate the creation of digital and green business models and enhance decision-making through improved traceability and lifecycle data. Twelve use cases were identified along the battery value chain, with three undergoing initial quantitative assessment: residual value estimation, recycling efficiency, and end-of-life collection.
Policy support needs	Effective regulatory frameworks are crucial to minimize implementation costs and aid adoption, especially for SMEs. Additional policy measures beyond current EU requirements are suggested to fully realize the passport's potential.
Consumer engagement	Battery Passports could enable consumers to make more informed purchasing and resale decisions, provided communication strategies are clear and accessible.

Source: Adapted from Battery Pass Consortium [23].

3.2 Business Models for Implementation

The implementation of the Battery Passport system offers a wealth of opportunities for both generating revenue and providing significant value to various stakeholders, including manufacturers, consumers, recyclers, and regulators. Given its broad scope and the complex nature of the battery industry, multiple business models can be employed to ensure the system's financial sustainability while delivering essential services that improve battery lifecycle management and compliance with regulatory standards. This section explores several potential business models for implementing the Battery Passport, each offering different approaches to revenue generation while supporting the goals of sustainability and transparency.

3.2.1 Subscription Model

In a subscription-based approach, stakeholders such as manufacturers or recyclers would pay a recurring fee to access the Battery Passport system. This model provides a steady revenue stream for platform operators and ensures financial sustainability. Stakeholders would benefit from ongoing access to real-time data and insights regarding battery lifecycle management, enabling them to optimize processes and comply with sustainability requirements.

Example 3.1 Similar to Software-as-a-Service (SaaS) models like Adobe Creative Cloud or Microsoft Office 365, users would pay a subscription to access the Battery Passport's features, such as data analytics and compliance tracking tools [24].

3.2.2 Pay-Per-Use Model

The pay-per-use model charges stakeholders each time they access specific data, making it suitable for smaller companies that may not require continuous access. This approach provides flexibility for businesses to scale their data usage based on need, ensuring that costs align with actual usage.

Example 3.2 Similar to Amazon Web Services (AWS), where users are billed for the computing power or data they consume, the Battery Passport could implement a usage-based fee structure. Companies could be charged based on the volume of battery data accessed or processed, offering a cost-effective solution for businesses with fluctuating data needs [25].

3.2.3 Integrated Service Offerings

The Battery Passport system can be bundled with value-added services, such as sustainability consulting, data analytics, or predictive maintenance tools. These integrated services can enhance the value proposition for stakeholders by offering a comprehensive suite of solutions tailored to battery management and regulatory compliance.

Example 3.3 Much like LinkedIn Premium, which provides job search tools, candidate recruitment, and data-driven insights, the Battery Passport could offer packages that include battery performance optimization, lifecycle management, and regulatory reporting services [26].

3.2.4 Data Monetization

The Battery Passport will collect valuable data on battery performance, material sourcing, and recycling processes. This anonymized data could be monetized by selling insights to third parties, such as market research firms, academic institutions, or governmental bodies interested in sustainability trends. By sharing anonymized lifecycle data, stakeholders can gain insights into market trends and emerging opportunities in battery technology and recycling.

Example 3.4 Similar to Mastercard Advisors, which monetizes transaction data to provide insights to financial institutions, the Battery Passport system could generate revenue by selling sustainability-related data to research entities and policymakers [27].

3.2.5 Public–Private Partnerships (PPPs)

PPPs could be utilized to fund and implement the Battery Passport system. Governments may partner with private companies to share costs, risks, and expertise, particularly for large-scale projects aimed at fostering sustainable practices. This model helps distribute the financial burden and ensures broader industry adoption by leveraging public funds for infrastructure development.

Example 3.5 The Gateway Arch Park Foundation and Indiana Toll Road are examples of successful PPPs that have raised funds through private investment and public sector involvement, showcasing the potential for shared benefits in large infrastructure projects [28] (Table 3.3).

The Battery Passport system presents a wide array of potential business models that can cater to diverse stakeholders while generating sustainable revenue

Table 3.3 Business models for the implementation and commercialization of Battery Passport systems.

Business model	Description	Example
Subscription model	Stakeholders pay a recurring fee for continuous access to the Battery Passport system, ensuring a steady revenue stream and ongoing access to real-time data and insights.	Similar to Adobe Creative Cloud or Microsoft Office 365, where users pay a subscription for access to software features and updates.
Pay-per-use model	Users are charged based on their usage of specific data, making it suitable for smaller companies with fluctuating data needs.	Like Amazon Web Services (AWS), where users pay for the computing power or data they consume, offering flexibility and cost-effectiveness.
Integrated service offerings	The Battery Passport is bundled with additional services such as sustainability consulting, data analytics, or predictive maintenance tools, providing a comprehensive suite of solutions.	Similar to LinkedIn Premium, which offers job search tools, recruitment services, and data-driven insights as part of a package.
Data monetization	Anonymized data collected by the Battery Passport on battery performance, material sourcing, and recycling processes is sold to third parties, providing insights into sustainability trends.	Like Mastercard Advisors, which monetizes transaction data to offer insights to financial institutions and other entities.
Public–private partnerships (PPPs)	Governments and private companies collaborate to share costs, risks, and expertise, fostering sustainable practices and ensuring broader industry adoption.	Examples include the Gateway Arch Park Foundation and Indiana Toll Road, where public and private sectors collaborate on infrastructure projects.

Table 3.4 Comparison of business models between base metals and energy transition minerals.

	Base metals	**Energy transition minerals**
Examples	Copper, aluminum, zinc	Lithium, cobalt, nickel (chemical)
Industry maturity	Well-established, mature infrastructure	Emerging industry with growing infrastructure
Feedstock	High material concentration but significant variance in alloy content	Lower concentration of materials due to complex, heterogeneous chemical compounds
Regulatory landscape	Relatively well-established regulations and policies, especially for steel and copper	Evolving regulations and policies to support recycling development
Market dynamics	Stable and well-established markets for recycled materials	Emerging and growing markets for recycled materials
Process complexity	Complex but well-understood metallurgical processes	Often more complex and customized given greater diversity of feedstock
Pretreatment requirements	Physical separation techniques, such as sorting, shredding, manual dismantling, and magnetic separation, are all commonly employed	
Processing	Pyrometallurgical and hybrid processes, particularly smelting, are often used for base metal recycling	Hydrometallurgy enables the selective leaching, extraction, and purification of the target metals from complex waste streams, but pyrometallurgy and hybrid processes are also used
Future processing methods	Recycling processes are often well-established and optimized	Growing focus on process intensification and efficiency improvements, including via the development of novel leaching agents, advanced separation techniques, and automation

Source: Recycling of Critical Minerals IEA / IEA / CC BY 4.0.

streams. Whether through subscription-based access, pay-per-use flexibility, integrated service offerings, data monetization, or PPPs, each model provides unique advantages tailored to the specific needs of manufacturers, recyclers, and regulators. As the system evolves, adopting one or a combination of these models can ensure its financial sustainability while driving the adoption of more transparent, efficient, and sustainable battery management practices. By leveraging these business models, the Battery Passport system will be positioned to play a key role in shaping the future of the battery industry, promoting sustainability, and facilitating regulatory compliance on a global scale (Table 3.4).

3.3 Financial Incentives and Support Mechanisms

3.3.1 Overview of Tax Incentives

The implementation of the Battery Passport and its related requirements is going to be extremely expensive for manufacturers, and in turn for consumers too. Governmental help will be essential for a timely and effective rollout of these extensive requirements. Tax incentives are a widely used tool to encourage companies to adopt sustainable practices and, in turn, for the implementation of Battery Passport systems. Now we will discuss a number of ways the governments can facilitate this rollout with economic incentives.

3.3.2 Designing Effective Tax Incentives

To maximize their impact, tax incentive programs should be tailored to the needs of different stakeholders, with clear eligibility criteria and well-defined goals. For instance, providing tax reductions for companies that achieve specific milestones, such as recycling targets or emissions reporting compliance, can drive broader adoption of the Battery Passport system.

3.3.3 Tax Credits and Subsidies

Many governments, including those in the EU, provide tax credits and subsidies to companies that comply with sustainability initiatives, such as the Battery Passport system. These incentives are designed to alleviate the financial burden on businesses, particularly small- and medium-sized enterprises (SMEs), which may face challenges in covering compliance costs.

3.3.4 Subsidies to Promote Battery Passport Adoption

Governments in regions like China, the European Union, and North America could be using subsidies to reduce the financial burden of adopting Battery Passports, particularly for SMEs. These subsidies could cover the costs of integrating digital systems, upgrading infrastructure, and meeting regulatory requirements. In China, subsidies have proven effective in accelerating the adoption of battery recycling and tracking technologies, fostering a more sustainable battery ecosystem. By promoting information sharing between battery manufacturers and formal recyclers, these subsidies help increase the volume of batteries recycled through compliant channels, thereby curbing the dominance of informal recyclers and enhancing environmental sustainability.

3.3.5 Grants and Funding Mechanisms

The EU and international organizations provide grants and funding to support research and development (R&D) in battery technologies and lifecycle management.

Grants are particularly common in high-growth industries such as EVs. These funds are often directed toward research and development (R&D), pilot projects, and scaling up new technologies. For example, grants can support the development of advanced digital platforms for lifecycle tracking, enabling companies to meet regulatory requirements efficiently.

3.3.6 Impact of Government Grants on Battery Adoption

Well-targeted grants can significantly impact the adoption of Battery Passports. By funding initiatives in areas like material recovery technologies, emissions tracking, and battery durability testing, grants help companies align with sustainability goals and regulatory standards. Moreover, grants often encourage cross-sector collaboration, fostering innovation and accelerating the integration of cutting-edge solutions into the battery supply chain. Government grants play a crucial role in accelerating the adoption of battery technologies by supporting initiatives in material recovery, emissions tracking, and battery durability testing. These grants not only help companies align with sustainability goals and regulatory standards but also foster cross-sector collaboration, driving innovation and integrating cutting-edge solutions into the battery supply chain.

For example, the Battery Pass Initiative has already received funding from the German Federal Ministry for Economic Affairs and Climate Action by resolution of the German Bundestag under grant agreement No. BZF335.

3.3.7 Public–Private Partnerships (PPPs)

Governments may establish PPPs to share the financial and operational responsibilities of implementing the Battery Passport system. These collaborations allow for public funds to support the development of digital infrastructure while private companies contribute their technical expertise and industry knowledge.

3.3.8 Risks that Support Mechanisms Must Address

While the Battery Passport system holds great promise for improving sustainability and transparency, it also presents several risks that could affect consumers and end users. The integration of new technologies may increase production costs, potentially limiting adoption in price-conscious industry segments. Additionally, consumers are increasingly concerned with data privacy and security. The collection and processing of sensitive information could hinder consumer trust and engagement. Moreover, if companies manipulate or falsify supply chain data, it creates serious trust issues that undermine the entire system, making it ineffective at promoting truly responsible sourcing. To mitigate these risks, it will be crucial for the Battery Passport system to ensure robust data privacy protections, cybersecurity measures, and transparency in data management. Equally important are financial incentives and support mechanisms that make compliance both feasible and attractive, especially for smaller manufacturers or companies operating on tight margins. Without adequate funding, technical support, or policy-driven incentives, companies may cut corners or manipulate data to avoid added costs, further eroding trust in the system.

Targeted subsidies, tax credits, or public–private funding programs can help offset the initial costs of integration, encourage ethical participation, and lower the barrier to entry for secure, verifiable data practices. These mechanisms must be designed not only to accelerate adoption but also to reinforce the integrity of the system, ensuring that Battery Passports genuinely drive responsible sourcing, consumer trust, and meaningful sustainability outcomes.

3.4 Global Trade

The Battery Passport system is gaining traction, with global trade policies and consumer behavior shifts being crucial to its success. This system aims to enhance sustainability, traceability, and transparency in the battery industry. However, it faces diverse challenges and opportunities due to international trade regulations, intellectual property protections, and changing consumer preferences for sustainable products.

Key factors influencing the adoption of the Battery Passport include nontariff barriers and trade regulations, which affect its implementation across regions. This section examines the economic impact of global trade policy changes and the role of intellectual property in fostering technological innovation. It also explores the potential for harmonizing global trade policies to support the widespread adoption of the Battery Passport system.

3.4.1 Nontariff Barriers and Trade Regulations

Nontariff barriers, such as technical standards, environmental regulations, and licensing requirements, play a pivotal role in determining how the Battery Passport can be implemented and adopted globally. The battery value chain extends across the entire world, often spanning jurisdictions with markedly different regulatory landscapes.

Trade between China, the United States, and the European Union is economically critical – and more uncertain than ever. The EU is adopting protectionist policies to shelter its automotive industry from external competition, and (as of the time of writing) the Trump administration is engaging in worldwide trade wars. Time will tell if these policies will be an effective lifeline to European and American manufacturing, as intended.

3.4.2 Economic Impact of Nontariff Barriers

Nontariff barriers can restrict the flow of batteries between regions, especially if regulations are not aligned. Batteries meeting the stringent environmental and technical standards set by the European Union (EU) may face barriers in non-EU markets if those markets do not adopt similar transparency and traceability requirements. This includes batteries produced within the EU as well as those manufactured elsewhere with EU consumers in mind. This fragmentation can increase compliance costs for manufacturers, impacting their competitiveness in international markets. Cross-border trade in batteries faces significant regulatory fragmentation, as different jurisdictions maintain varying definitions of waste and recycling standards that create barriers to international battery lifecycle management [29].

3.4.3 Intellectual Property Considerations

Implementing the Battery Passport requires sophisticated technologies. Strong IP protections are essential for safeguarding these technologies and encouraging investment in regions adopting the Battery Passport. For example, a China-led passport initiative could sell their IP and technology for profits in the future. Countries with weaker IP laws may struggle to attract necessary technology developers and investors, potentially slowing the system's adoption [30].

3.4.4 Coordinating Global Trade Policies

There is a clear need for coordinated global trade policies, as disparate regulations can hinder the widespread adoption of the Battery Passport. The absence of unified standards may result in a fragmented system, complicating efforts to standardize battery supply chain transparency.

3.4.5 Opportunities for Coordination

Coordinated trade policies can help create a more unified framework for implementing the Battery Passport across regions. For example, aligning standards between major markets could reduce compliance costs and facilitate trade. By integrating sustainability requirements into trade agreements, governments can incentivize companies to adhere to common standards, promoting a more streamlined approach to battery regulation [31].

3.4.6 The Role of Trade Agreements in Promoting Adoption

Trade agreements have increasingly become tools for promoting sustainable practices in various industries, including battery manufacturing. By incorporating sustainability and traceability clauses into trade agreements, policymakers can encourage the adoption of the Battery Passport system.

Trade agreements, such as those led by the EU, demonstrate how sustainability clauses can be integrated into international agreements to create market incentives for adoption, ultimately promoting global sustainability and transparency in the battery industry. Through these policy efforts, stakeholders can pave the way for the widespread implementation of the Battery Passport system, driving more sustainable practices and ensuring greater accountability in battery supply chains worldwide (Figure 3.10).

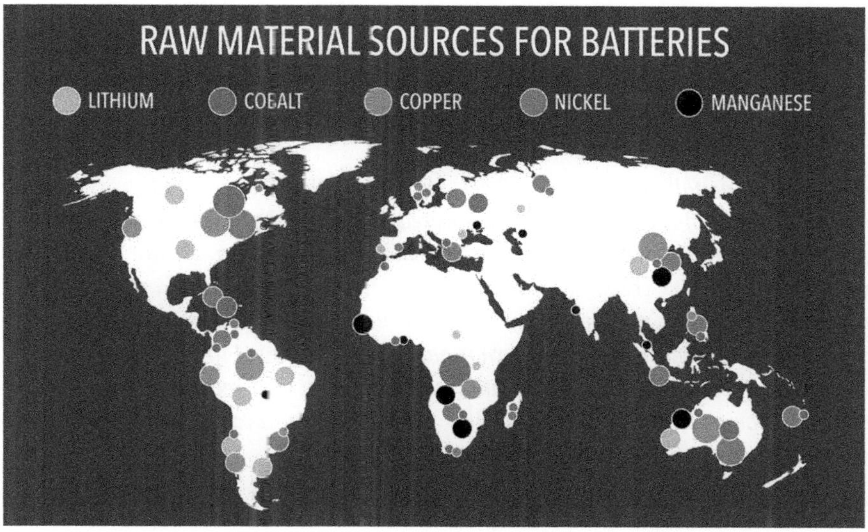

Figure 3.10 Raw Material Sources for Batteries. *Source:* [32] / Canadian Auto Dealer.

3.4.7 Manufacturer's Economic Impact

For manufacturers, fragmented trade regulations and inconsistent standards across borders translate directly into higher compliance costs and limited market access. Companies operating in multiple jurisdictions must adapt production processes to meet varying environmental and traceability requirements, which increases overhead and slows market expansion.

3.4.8 Consumer's Economic Impact

As manufacturers incur higher costs to comply with multiple regulatory regimes, these expenses may be passed on to consumers in the form of higher product prices, especially for EVs and energy storage systems. Moreover, the lack of harmonized trade standards could result in restricted product availability across regions, limiting consumer choice or delaying the rollout of sustainable battery-enabled products. If trust in sustainability claims varies between regions due to weak oversight or inconsistent enforcement, consumers may also experience uncertainty around the validity of environmental certifications, undermining their willingness to pay a premium for greener alternatives.

3.4.9 Government's Economic Impact

Governments that fail to coordinate their regulations with major trading partners risk isolating domestic industries and discouraging foreign investment in local battery ecosystems. Without alignment, domestic manufacturers may face trade disadvantages, and export opportunities could shrink. Additionally, inconsistent IP enforcement may deter global tech firms from deploying solutions in certain markets, stalling innovation and weakening regulatory credibility. On the other hand, governments that integrate Passport standards into trade agreements could gain strategic leverage by positioning their industries at the forefront of sustainable, compliant global trade.

3.4.10 Recycling Companies' Economic Impact

For recyclers, nonaligned trade policies present a direct barrier to cross-border movement of end-of-life batteries and recovered materials. Stringent or incompatible export/import rules for used batteries and critical minerals can disrupt operations and drive up logistics costs. Without globally consistent definitions and standards for material recovery, recyclers may also face uncertainty in proving compliance, reducing their competitiveness in international supply chains. Furthermore, weak IP protections in some countries may inhibit access to advanced recycling technologies, increasing operational risk and slowing investment in next-generation recovery systems (Figures 3.11–3.13) (Table 3.5).

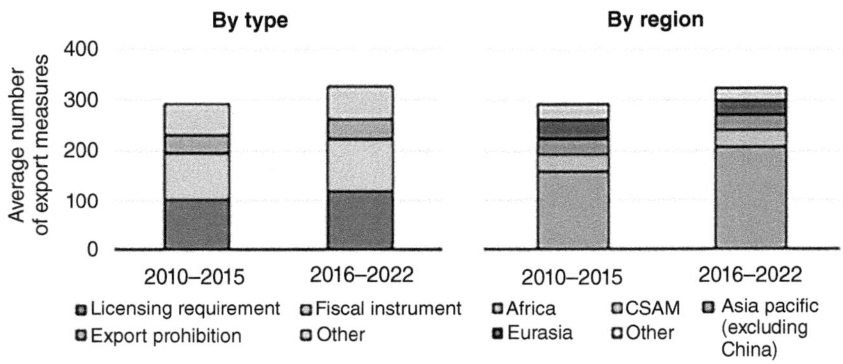

Figure 3.11 Export control measures for critical minerals and metals waste and scrap. *Notes:* CSAM, Central and South America. Other regions include Europe. China, the Middle East, and North America. Covered minerals and metals include copper, nickel, aluminum, molybdenum, cobalt, and manganese. Only includes measures that have not ended. Fiscal instruments include export taxes, fiscal tax on exports, value-added tax rebate reduction/withdrawal, and export surtax. *Source:* Recycling of Critical Minerals IEA / IEA / CC BY 4.0.

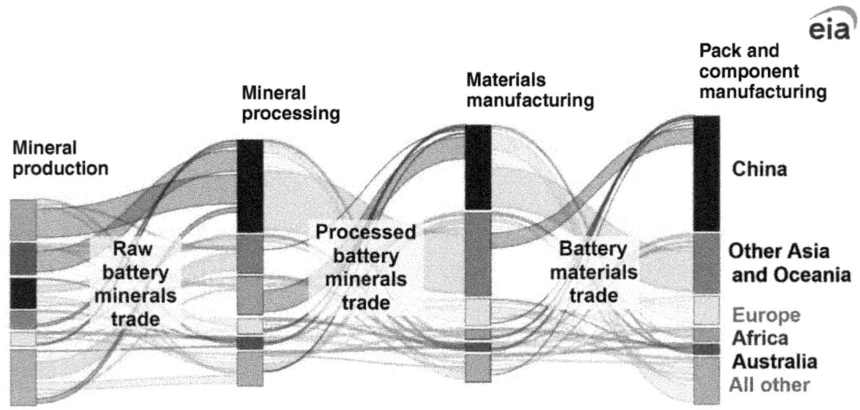

Figure 3.12 World battery minerals and materials trade by region. *Source:* U.S. Energy Information Administration (EIA) / Public Domain.

Table 3.5 Different barriers for trade.

Barrier type	Description	Examples	Typical impacts
Technical regulations	National standards that differ from international norms	Unique safety, labeling, or testing standards	Increased compliance costs; exclusion of foreign goods
Licensing requirements	Government authorization needed to import/export	Quotas, import licenses, or dual-use export controls	Reduced market access; potential delays
Customs procedures and delays	Bureaucratic procedures causing inefficiencies	Lengthy documentation, arbitrary inspections, lack of digital systems	Increased time and cost of cross-border trade
Local content requirements	Mandates to use a certain percentage of local inputs	EVs must include a percentage of locally sourced batteries	Discriminates against foreign suppliers; reduces competitiveness
Subsidies and state aid	Government support favoring domestic firms	Energy subsidies, tax rebates for local producers	Distorts market competition; violates WTO principles if discriminatory
Data localization laws	Requirement that data be stored/processed domestically	EU GDPR restrictions; China's data sovereignty rules	Raises IT and compliance costs for foreign firms
Intellectual property restrictions	Limited protection/enforcement of IP rights	Patent registration delays; forced IP transfer	Reduces incentives for innovation; deters high-tech investment
Mutual recognition failures	Lack of acceptance of foreign conformity assessments	The EU not recognizing certifications from third countries	Duplicate testing/certification increases cost
Environmental or carbon standards	Varying eco-labels, carbon border taxes, or green procurement rules	EU Carbon Border Adjustment Mechanism (CBAM)	Disadvantages firms in countries with weaker environmental policies
investment restrictions	Barriers to foreign ownership or control	Caps on foreign equity in strategic sectors	Limits FDI and cross-border control
Digital trade restrictions	Barriers to e-commerce and digital services	Censorship, restrictions on cross-border payments, blocking platforms	Limits digital market access; affects trade in services

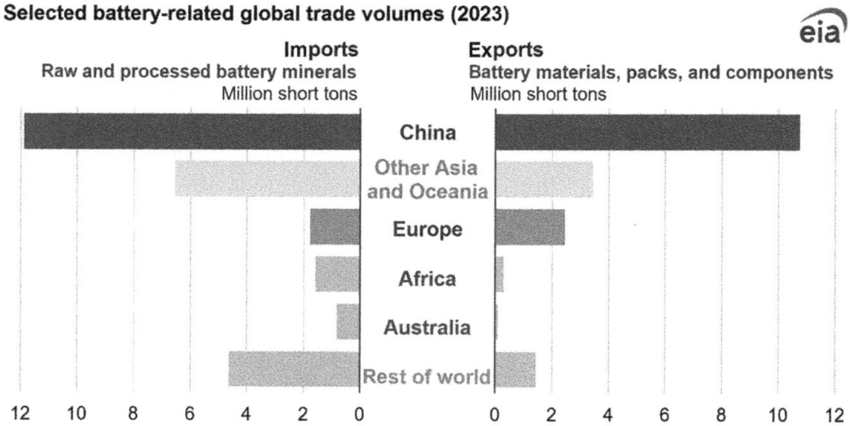

Figure 3.13 Selected battery-related global trade volumes (2023). *Source:* U.S. Energy Information Administration (EIA) / Public Domain.

3.5 Economic Risks for Stakeholders

3.5.1 Creating New Economic Risks

See Table 3.6.

Table 3.6 Investment risk related to batteries.

Key risk	Applicability	Description
Regulatory	Both	Lack of a clear and stable regulatory environment, including the absence of clear policy signals, manufacturing support, equal access to power markets for battery storage, and a strategy for the remuneration of services provided.
Financing and affordability	EV batteries	Ability to afford the high upfront costs of EVs and higher financing costs due to higher interest rates and limited access to debt financing, especially in emerging markets and developing economies.
Framework conditions	EV batteries	Absence of adequate EV charging infrastructure and lack of proven business models.
Remuneration	Battery storage	Delayed payments or under-recovery from distribution companies for energy storage services provided.

Source: Batteries and Secure Energy Transitions (IEA) / IEA / CC BY 4.0.

While government policies and financial incentives can help mitigate certain risks associated with the implementation of the Battery Passport system, they also introduce new economic challenges for stakeholders in the ecosystem. These risks, such as over-regulation, regulatory uncertainty, and dependency on government incentives, could impact the ability of companies to operate efficiently and remain competitive. This is especially true for the European battery industry, which currently struggles to compete with its Chinese counterpart. The growing complexity of compliance requirements, combined with the instability of evolving regulations, may affect market participation, innovation, and long-term business strategies. This section explores how these new economic risks can influence the Battery Passport system and offers potential strategies to mitigate their impact.

3.5.1.1 Over-Regulation

The implementation of stringent regulations could result in over-regulation, where businesses face excessive compliance requirements. This could increase operational complexity and elevate costs, particularly for companies already dealing with tight margins, like in the battery industry. The risk of over-regulation may discourage smaller companies from participating in the Battery Passport system and the battery industry as a whole, leading to market consolidation where only large firms with sufficient resources – most of which are based outside of Europe – can remain competitive.

In other words, over-regulation can stifle innovation and deter new market entrants, further weakening European industry. Smaller firms may find it financially unfeasible to comply with complex regulations, resulting in a less diverse market dominated by larger companies.

3.5.1.2 Regulatory Uncertainty

In regions where battery regulations are still evolving, there is a risk of regulatory uncertainty. Changes in government policies or delays in the implementation of regulations can create an unstable environment for businesses, making it challenging to plan long-term investments in technology and infrastructure.

Mitigating Risks: Governments can reduce regulatory uncertainty by providing clear, long-term roadmaps for implementing Battery Passport systems. Consistent enforcement and transparent guidelines will help businesses adapt more confidently to new requirements.

3.5.1.3 Dependency on Government Incentives

Relying heavily on government subsidies and tax incentives could create financial dependencies for businesses. If these incentives are reduced or withdrawn in the future, companies may struggle to maintain compliance with the Battery Passport system without the financial support they previously relied on.

Mitigating Risks: Businesses could develop strategies that do not solely rely on government incentives, although, in many cases, it is difficult to do. By investing in cost-effective, scalable, and flexible compliance solutions, companies can prepare for

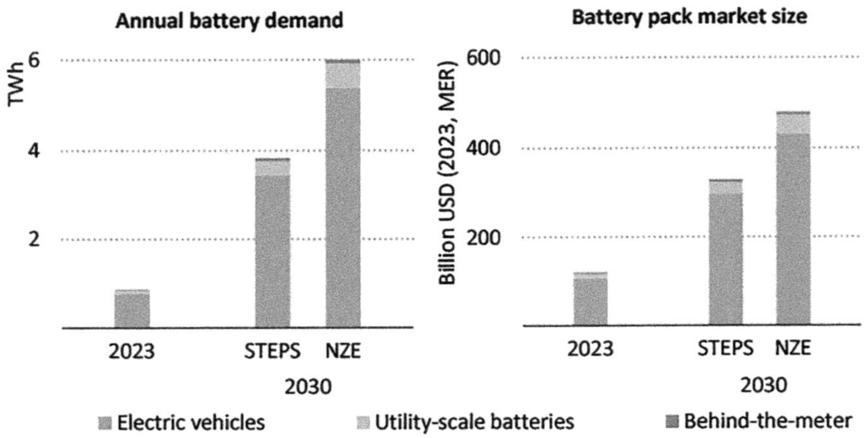

Figure 3.14 Annual battery demand and battery pack market size by application and scenario, 2023 and 2030. *EVs account for about 90% of battery demand to 2030, while the value of the battery pack market reaches USD 330 billion in the STEPS and nearly USD 500 billion in the NZE Scenario. Source:* Batteries and Secure Energy Transitions (IEA) / IEA / CC BY 4.0.

potential changes in government policy and reduce dependency on external financial support.

Many companies, particularly SMEs, may simply not have the financial capacity to absorb the extensive costs associated with implementing the Battery Passport system. In sectors already under pressure from global competition, fluctuating raw material costs, and supply chain disruptions, the addition of mandatory DPP infrastructure could push many firms beyond their financial limits. While larger corporations might be able to treat these costs as strategic investments, smaller players may face the stark choice of risking noncompliance or exiting the market altogether. This creates the risk of a two-speed industry, in which only the most capitalized firms can afford to comply, thereby widening inequality and consolidating market power. If these dynamics are not addressed through scalable compliance pathways or tiered regulatory obligations, the Battery Passport could unintentionally erect barriers to entry and innovation, undermining the system's goal of fostering broad-based sustainability across the battery value chain (Figure 3.14).

3.5.2 Risk Analysis for Manufacturers

The implementation of the Battery Passport system introduces various economic risks for manufacturers, particularly related to the added financial burden of new technologies, regulatory compliance, and market resistance. While the system aims to enhance transparency and sustainability, manufacturers face significant challenges in adapting to these requirements, especially smaller companies with limited resources. These risks are not only financial but also strategic, as they could affect a manufacturer's market position, reputation, and ability to attract investment. This section explores the specific economic risks manufacturers may encounter with the introduction of the Battery Passport system, from increased production costs to compliance with evolving regulatory frameworks.

3.5.2.1 Increased Production Costs

Manufacturers must invest in systems to track, collect, and manage data across the entire battery lifecycle. These new requirements, including the need for transparency in material sourcing and compliance with sustainability metrics, are expected to increase production costs. Smaller manufacturers may struggle with the financial burden of upgrading their processes to meet these standards, which include the integration of advanced tracking systems and sustainable manufacturing practices [1]. While large manufacturers may be able to absorb or offset these costs, smaller players face margin erosion, cash flow strain, or even market exit. Without regulatory support or cost-sharing mechanisms, the added financial burden could discourage innovation or shift production away from sustainability altogether.

3.5.2.2 Infrastructure and Capital Risks

As explained above, the economic risks associated with infrastructure requirements for the Battery Passport system are substantial and the primary reason for increased production costs. Implementing the necessary infrastructure, including IoT-enabled production facilities, data centers, and integrated networks, requires significant capital expenditure. Manufacturers must either upgrade existing facilities or construct entirely new ones to meet the technical requirements of real-time data tracking and lifecycle transparency. These upgrades demand major capital investment with uncertain (or nonexistent) return on investment. For many manufacturers, especially in developing economies, the lack of regional digital infrastructure or financing options compounds the challenge. Worse, these investments may be rendered obsolete if global standards shift or if new regulatory demands require further adaptation, turning infrastructure into a stranded asset. This particularly impacts smaller firms, such as most EU-based manufacturers, who may lack the resources to make these investments.

Additionally, retrofitting existing production lines or supply chain infrastructure can lead to operational downtime, increasing costs and delaying production schedules. These risks are compounded by the rapid pace of technological innovation, which may render some infrastructure investments obsolete before manufacturers can fully recoup their costs. This creates significant barriers to entry, increasing the advantage of current industry giants, most of whom operate in China.

Furthermore, infrastructure investments may disproportionately affect manufacturers in regions with underdeveloped digital ecosystems. This imbalance risks widening the competitive gap between manufacturers in developed and developing markets, further complicating the economic feasibility of adopting the Battery Passport system across the global industry. Additionally, the associated costs, along with the need to hire skilled personnel to manage these systems, could be prohibitive, especially for companies with limited resources [9].

The upfront costs are high, the talent pool is limited, and the risk of implementation failure is real. Misaligned tech investments, such as incompatible platforms or underutilized software, could drain resources without delivering compliance or performance benefits, especially in the uncertain global trade situation mentioned before. For smaller or mid-sized firms, the steep learning curve and cost of onboarding tech talent create a high barrier to entry. Large companies, therefore, are not only able to absorb the upfront costs but also stand to benefit from the reduction in competition.

3.5.2.3 Market Resistance

The Battery Passport system's emphasis on transparency and sustainability may lead to higher battery prices. Manufacturers will likely pass the costs of implementing the system down to consumers, which could result in market resistance, particularly in price-sensitive sectors. Consumers may be unwilling to pay the premium for sustainability features if they do not perceive an immediate benefit or if lower-cost alternatives are available.

This cost transfer along the value chain risks higher prices for consumers, potentially leading to reduced sales or slower adoption in competitive markets. If compliant batteries become too expensive for mass-market consumers, there could be a shift toward lower-cost, less sustainable alternatives, thereby undermining the environmental goals of the Battery Passport system (Figure 3.15).

3.5.2.4 Regulatory Compliance

Manufacturers must continuously adapt to changing regulations to ensure compliance. This may require significant resources, particularly in regions like the EU and China, where regulations are stringent and ever-evolving. Failure to comply could lead to significant penalties and reputational damage. Manufacturers must also stay informed of regulatory changes to avoid noncompliance risks [34]. The automotive industry has previously faced significant challenges related to regulatory compliance, particularly concerning emissions standards and safety regulations. For instance, there have been notable cases where automobile manufacturers failed to meet regulatory requirements, leading to substantial penalties and damage to their reputations.

One prominent example is the "Dieselgate" scandal involving a major automobile manufacturer. The company was found to have installed software in diesel engines that could detect when they were being tested, changing the performance to improve results. During normal driving conditions, these vehicles emitted pollutants up to 40 times above the legal limits. This deception led to massive fines, legal actions, and a severe blow to the company's reputation.

This example underscores the importance of compliance with regulations and the potential consequences of failing to do so. Similarly, manufacturers in the battery industry must ensure that they adhere to evolving regulations to avoid penalties and maintain their market standing.

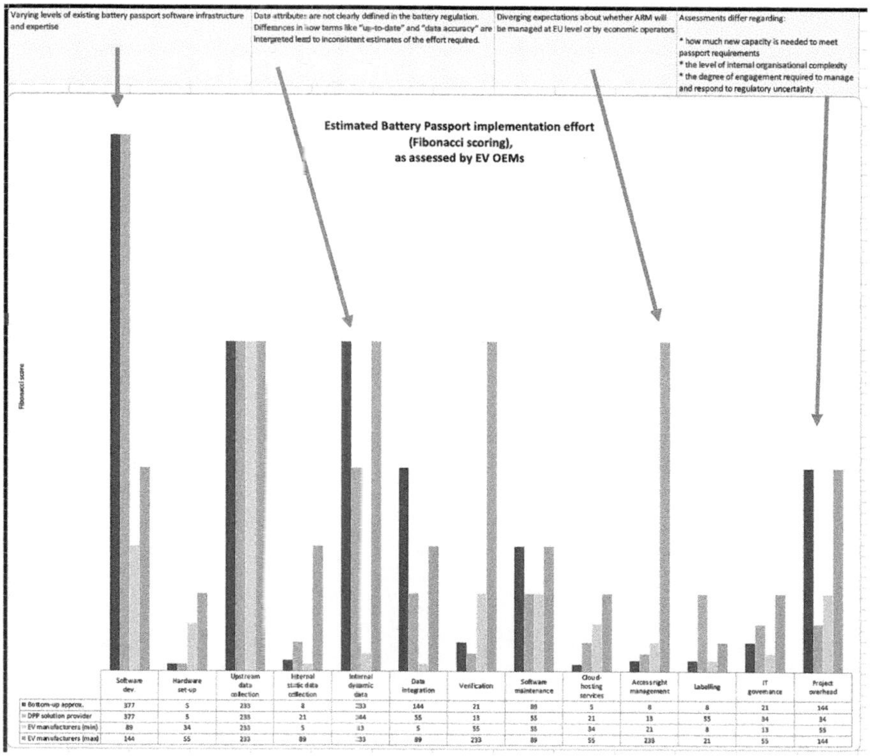

Figure 3.15 Battery Passport implementation efforts (Fibonacci scores) as approximated by EV OEMs. *Source:* Data from Battery Pass Consortium. https://thebatterypass.eu/assets/images/value-assessment/pdf/2024_BatteryPassport_Value_Assessment_Synthesis.pdf (accessed 15 December 2026).

The pace and complexity of regulatory change pose serious risks for manufacturers. Battery Passport compliance is not static; it requires data updates, multijurisdictional reporting, and proactive adaptation to new rules. Even minor lapses in data or documentation could result in fines, exclusion from markets, or reputational damage. For companies with global operations, keeping up with evolving compliance requirements in the EU, China, and North America represents a continuous operational and legal challenge (Figure 3.16).

3.5.2.5 Supply Chain Risks

The Battery Passport requires due diligence across the entire supply chain to ensure compliance with social and environmental standards. This increases the risk for manufacturers, as they are held accountable for unethical practices or environmental violations by their suppliers. Failure to identify and mitigate these risks could result in legal action or consumer backlash, further jeopardizing brand reputation and market position [35]. The system's transparency means that any unethical mining, labor abuses, or environmental harm by suppliers can indirectly affect the manufacturers using their materials. Consequently, manufacturers must

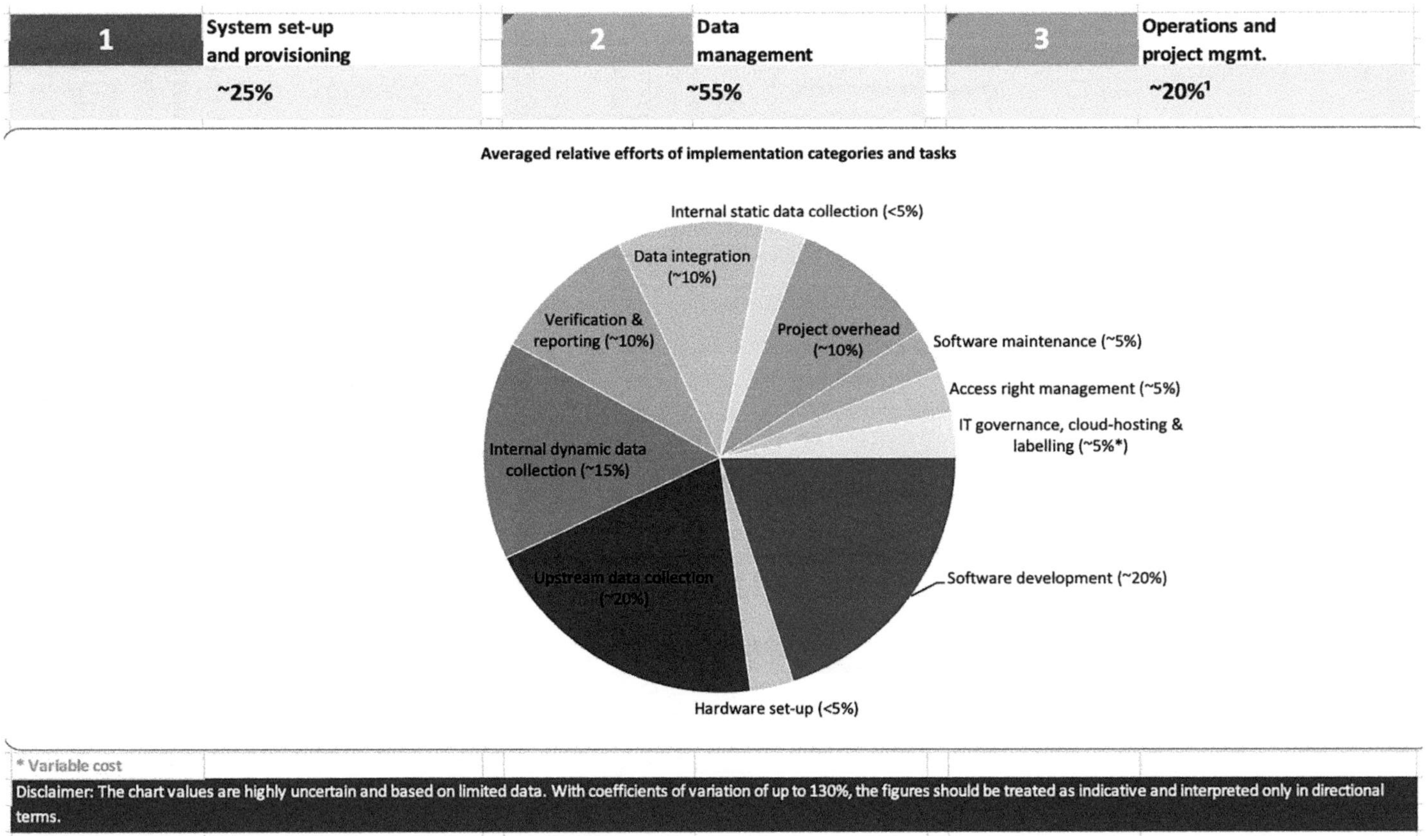

Figure 3.16 Averaged relative efforts of implementation categories and tasks.[1] **Disclaimer:** Values in the chart are highly uncertain and based on limited data, with coefficients of variation[2] of up to 130% → numbers should only be interpreted directionally. *Source:* Data from Battery Pass Consortium. https://thebatterypass.eu/assets/images/value-assessment/pdf/2024_BatteryPassport_Value_Assessment_Synthesis.pdf (accessed 15 December 2026).

implement extensive due diligence, which adds cost and operational complexity. A single failure in the supply chain could lead to public backlash or regulatory penalties, eroding trust and market access.

The Battery Passport is designed to increase pressure across the entire supply chain to comply with ESG standards. Increased transparency across a battery's lifecycle means that manufacturers face greater accountability for the actions of their upstream suppliers, whether directly or indirectly. In other words, if a supplier is found to be engaging in unethical or environmentally irresponsible practices, manufacturers using their materials may face legal action or consumer backlash that could threaten their market position.

This increases the importance of due diligence so that companies can identify and mitigate these legal and reputational risks. This adds additional costs and operational complexity across the supply chain; a single failure at any point could trigger public backlash or regulatory penalties, eroding trust and market access.

3.5.2.6 ESG Compliance Risks

As the Battery Passport facilitates transparency in supply chains, it will become easier for investors to assess a company's ESG compliance. Manufacturers that fail to meet stringent ESG standards may face difficulties in securing investments, lose market share, or suffer reputational damage. ESG performance is becoming a critical factor for financiers, and noncompliance could have long-term financial consequences [12]. This further increases intermanufacturer competition.

As ESG metrics become embedded in investment and procurement decisions, Battery Passports make it easier for external stakeholders to audit performance and flag noncompliance. Manufacturers that fall short on emissions, sourcing ethics, or labor standards risk being downgraded by investors, dropped by customers, or blacklisted by regulators. Unlike traditional supply chain performance issues, ESG failures carry lasting reputational and financial consequences, which can impact access to capital, insurance, and growth opportunities. These factors further increase investment risks and, in turn, increase the costs of financing one's operations (Table 3.7).

The Battery Passport system presents substantial economic risks for manufacturers, particularly in terms of the significant upfront investments required for new technologies and the potential resistance from price-sensitive markets. The costs of integrating advanced tracking and data management systems, along with ensuring compliance with stringent sustainability standards, could challenge many manufacturers, especially smaller players in the industry. Furthermore, the system's emphasis on transparency and ESG compliance introduces additional risks related to supply chain accountability, regulatory compliance, and reputational damage. While the Battery Passport system promotes sustainability, manufacturers must carefully navigate these risks to maintain competitiveness, secure investments, and avoid the potential fallout from noncompliance. Ensuring successful integration of the Battery Passport system will require strategic investments in technology, workforce development, and proactive risk management to capitalize on the long-term benefits while mitigating short-term economic challenges.

Table 3.7 Mitigation strategies.

Mitigation strategies	Details
Increased production costs	*Seek government subsidies, implement phased rollouts, and participate in industry consortiums to share costs.*
Infrastructure and capital risks	*Adopt modular infrastructure, explore shared or leased facilities, and partner with governments to develop regional digital ecosystems.*
Investment in new technologies	*Focus on cost-effective solutions, use open-source platforms, and prioritize workforce upskilling.*
Market resistance	*Educate consumers on sustainability benefits, offer tiered pricing models, and demonstrate added value by the Battery Passport.*
Regulatory compliance	*Establish compliance monitoring teams, automate regulatory tracking, and engage proactively with regulators for early adaptation.*
Supply chain risks	*Conduct supplier audits, diversify sourcing, and use transparency platforms to monitor and verify compliance.*
ESG compliance risks	*Communicate ESG efforts to build trust, secure third-party certifications, and align with investor priorities on sustainability.*

3.5.3 Risks for Consumers and End Users

3.5.3.1 Potential Increase in Product Costs

Implementing the Battery Passport system is likely to increase the prices of battery-powered products. The additional costs associated with data management, compliance with sustainability standards, and the integration of advanced tracking technologies may be passed down the value chain, leading to higher prices for consumers. This could reduce market adoption, especially in price-sensitive segments where the willingness to pay extra for sustainability features may be limited [1].

The Battery Passport introduces mandatory processes like traceability, emissions tracking, and data sharing that were previously optional or nonexistent. These functions require new tools, data infrastructure, and oversight, all of which drive up per-unit production costs. Without regulatory support or cost-sharing mechanisms, the added financial burden could discourage innovation or shift production away from sustainability altogether. This may lead to technological stagnation or cost-cutting measures that result in a decline in product quality.

Price sensitivity is a significant barrier in mass-market adoption of EVs and other related markets. In regions without incentives or subsidies, this could reduce accessibility and widen the affordability gap, slowing the transition to cleaner technologies. Consumers may also resist paying more for features they do not fully understand or perceive as immediately beneficial, risking market stagnation or the growth of gray-market alternatives. Overall, it is unclear whether the trade-off between better circularity and transparency and higher costs for consumers is a net positive.

3.5.3.2 Data Privacy Concerns

The Battery Passport system may eventually involve collecting and sharing detailed data throughout the battery lifecycle. While this data is valuable for stakeholders, it raises concerns about privacy, particularly regarding access and data-sharing policies. Consumers may be apprehensive about how much information about their battery usage is being shared with manufacturers, governments, or third parties. There are also risks related to compliance with data protection regulations, such as the General Data Protection Regulation (GDPR), which governs how personal data is processed and stored [36]. The rise of (perceived) higher privacy alternatives in many industries (like Telegram or Apple's privacy-centered marketing) confirms that this concern has an impact on consumer behavior.

If Battery Passports begin to capture granular data, ranging from battery performance and location history to usage patterns, they introduce new forms of data exposure for consumers. Even if the data is not directly personal, its aggregation and linkage to devices or vehicles can inadvertently reveal private behaviors. Without clear limits on how this information is collected, stored, and shared, consumers may feel surveilled or exploited. If public trust erodes, adoption may be undermined not by cost, but by a fear of losing control over one's digital footprint.

3.5.3.3 Data Security Risks

As the Battery Passport relies on digital systems for data storage and transfer, there are inherent risks of data breaches or unauthorized access. Consumers may be concerned about the security of sensitive information, especially if data is shared across interconnected networks. While blockchain technology can offer tamper-proof records, it is not entirely immune to cyberattacks. Therefore, the system's reliance on interconnected digital networks may expose it to security vulnerabilities that could compromise data integrity [9].

Reliance on interconnected systems IoT sensors, cloud platforms, and blockchain networks create a broad attack surface for hackers. If a breach were to expose consumer-linked data or allow unauthorized access to battery devices, the fallout could include financial loss, damaged property, or even safety risks (e.g., tampering with EV battery management systems). Even the *perception* of security vulnerabilities can discourage consumers from embracing smart battery-enabled products. A single high-profile breach could trigger widespread backlash, especially if trust is not proactively built into the system.

3.5.3.4 Trust Issues in Supply Chain Data

Concerns have been raised over the accuracy and transparency of data provided by some stakeholders in the supply chain, particularly mining companies that may not fully disclose all relevant information about their sourcing practices. If data transparency is not managed by an independent authority or trusted entity, consumers may remain skeptical about the validity of sustainability claims. Trust issues could undermine the effectiveness of the Battery Passport in promoting responsible sourcing and circular economy practices 4 [35].

The Battery Passport promises to certify ethical and sustainable sourcing, but that promise is only as strong as the weakest data contributor. If suppliers, especially those in high-risk regions, input inaccurate, incomplete, or misleading information, it threatens the credibility of the entire system. Consumers may grow skeptical of "green" or "ethical" claims if they sense the system lacks independent verification. This trust gap can lead to reputational damage for brands and reduced consumer engagement, ultimately weakening the market signal the Passport was designed to amplify.

Summary While the Battery Passport system offers substantial benefits in terms of sustainability and transparency, it also introduces key risks for consumers and end users that need to be addressed for its widespread acceptance. Potential increases in product costs, driven by the integration of new technologies and compliance with sustainability standards, may limit adoption, particularly in price-sensitive markets. Furthermore, data privacy and security concerns are critical, especially as detailed lifecycle data is shared among stakeholders across digital networks. To ensure consumer confidence, it will be essential to implement robust privacy protections and secure data management protocols. Finally, establishing trust in the accuracy and transparency of supply chain data is necessary to prevent skepticism and ensure the system's credibility. Addressing these concerns proactively will be key to achieving the full potential of the Battery Passport system while safeguarding the interests of consumers and end users.

3.5.4 Risks for Government Bodies

The implementation of the Battery Passport system also presents distinct economic risks for government bodies, which play critical roles in regulation, oversight, and incentivizing adoption. Governments face challenges related to the costs of developing and maintaining regulatory frameworks, investing in infrastructure to support the system, and managing the risks associated with uneven implementation across regions. These risks must be addressed to ensure the system's success and alignment with broader sustainability and circular economy goals (Figure 3.17).

3.5.4.1 Regulatory and Compliance Costs

The Battery Passport system's emphasis on transparency necessitates robust regulatory and compliance frameworks, which can incur substantial costs.

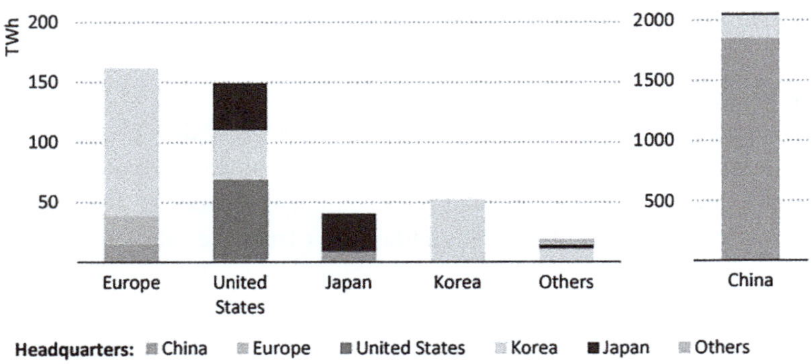

Figure 3.17 Lithium-ion battery manufacturing capacity by manufacturers' headquarter locations, 2023. *Battery producers in Korea lead as primary investors in overseas markets with over 350 GWh of manufacturing capacity, followed by Japan and China. Source:* Batteries and Secure Energy Transitions (IEA) / IEA / CC BY 4.0.

Governments are tasked with developing legal standards, auditing protocols, and reporting requirements to ensure the system's effectiveness. However, the return on investment for these efforts is uncertain.

A significant concern is the potential for these frameworks to become symbolic rather than effective, particularly if governments lack the necessary tools, personnel, or legal authority to enforce compliance. Without adequate enforcement, the Battery Passport system risks becoming ineffective, as bad actors could exploit weak oversight to falsify sustainability credentials. This not only erodes trust in the system but also places compliant stakeholders at a disadvantage. Furthermore, inadequate auditing and enforcement capacity could lead to reputational damage for governments, especially if sustainability violations are discovered postcertification.

3.5.4.2 Infrastructure Investments

The Battery Passport system relies on advanced infrastructure that will need to be financed somehow. If the private sector is unwilling or unable to shoulder this responsibility, it may fall on governments – potentially straining public budgets.

Governments are under increasing pressure to support Battery Passport functionality, which is emerging more slowly than expected. Rising support costs can divert funding from other critical infrastructure projects such as healthcare, transportation, or education.

3.5.4.3 Market Disruption Risks

The introduction of the Battery Passport system may disrupt existing markets by increasing compliance costs for manufacturers and other stakeholders. This could lead to resistance from industries or regions that are heavily dependent on battery production but are not well-prepared for the shift to transparency and sustainability. This would not be the first time this has happened in the European vehicle industry in the 21st century, as similar resistance arose with the introduction of emissions standards (e.g., Euro norms) and the shift toward electric mobility, which also imposed new compliance costs and disrupted traditional manufacturing practices. Governments may face backlash from stakeholders opposed to the economic implications of the system, complicating its implementation and adoption.

The Battery Passport introduces structural changes to how batteries are made, tracked, and traded. These shifts may heavily impact regions reliant on cost-competitive manufacturing, which may now face steep compliance costs. Resistance from local industries, labor groups, or regional governments could slow or stall implementation. Political leaders risk facing public or industrial backlash if the transition is perceived as top-down or if it disproportionately burdens small and medium-sized enterprises. Without careful stakeholder engagement, governments may find themselves navigating a politically charged implementation landscape.

3.5.4.4 Impact on Economic Value and Job Creation

The Battery Passport system is expected to contribute significantly to the creation of jobs across various sectors. By promoting a low-carbon economy, the system is poised to generate additional economic value within the renewable energy sector and encourage sustainable manufacturing practices [37]. However, in the context of other economic risks, this benefit may not be as straightforward as it appears.

While the Battery Passport system is designed to support sustainable growth and job creation, it may also introduce economic challenges if not carefully implemented. SMEs could struggle with the high compliance costs associated with data reporting, digital infrastructure, and regulatory alignment, potentially leading to reduced competitiveness or even market exit. Additionally, the automation of data collection and supply chain tracking through AI and blockchain may reduce demand for certain administrative and logistics jobs, creating displacement in the workforce.

Furthermore, if the system has a negative impact on small and medium-sized enterprises, whatever economic growth and job creation occurs will be concentrated in regions that already have advanced digital and industrial capabilities. Regions that are already playing catch-up, including Europe, risk falling even further behind. This geographic and technological divide could hinder global equity in value creation and slow down broader innovation, especially if startups and local recyclers are unable to meet strict requirements. Without targeted support mechanisms, such as funding for compliance or workforce retraining, the Battery Passport could inadvertently create economic exclusion rather than inclusive green growth, leading to fewer jobs in the field.

3.5.4.5 Mitigation Strategies

International Collaboration: *Governments can reduce costs by continuing to collaborate with international bodies such as the GBA to further develop standardized frameworks and share best practices. This approach ensures consistency while avoiding duplication of efforts.*

PPPs: *Partnering with private sector stakeholders, such as technology companies and manufacturers, can offset infrastructure and regulatory costs. These partnerships can foster innovation and share financial burdens.*

Incentivizing Adoption: *Governments can provide financial incentives, such as tax breaks or subsidies, to support stakeholders in adopting the Battery Passport system. These incentives can help mitigate resistance and promote widespread implementation.*

Capacity Building: *Investing in education and training programs for public sector employees and industry stakeholders will help build the expertise needed to manage and enforce the system effectively.*

Government bodies face significant economic risks in implementing the Battery Passport system, including regulatory costs, infrastructure investments, and the potential for uneven adoption. However, by leveraging collaboration, partnerships, and targeted incentives, governments can mitigate these risks and create a supportive environment for the system's success. Proactively addressing these challenges will be essential to ensuring the Battery Passport system achieves its sustainability and circular economy objectives while minimizing the economic burden on public institutions.

3.5.5 Risk Analysis for Recycling Companies

The implementation of the Battery Passport system introduces significant economic risks for recycling companies, particularly in terms of compliance, infrastructure investments, and market viability. As key stakeholders in the circular economy, recycling firms must adapt to increased transparency and data-sharing requirements while managing the financial and operational challenges associated with processing end-of-life batteries. These risks are not only financial but also strategic, as they could impact the ability of recycling companies to compete, expand, and remain profitable in a rapidly evolving regulatory landscape (Figures 3.18–3.21).

3.5.5.1 Increased Operational Costs

Recycling companies must invest in upgrading their facilities to comply with the tracking and transparency requirements of the Battery Passport system. This includes modernizing recycling processes to align with stricter environmental standards, integrating digital tracking technologies to ensure full traceability of recovered materials, and hiring skilled personnel to manage compliance. Smaller recycling firms may struggle with these additional financial burdens, limiting their ability to compete with larger, well-funded players. The cost of implementing advanced sorting and processing technologies, particularly for lithium-ion batteries, adds further pressure, as these investments may take years to generate a return.

Compliance with Battery Passport requirements will force recycling companies to invest heavily in digital traceability, material tracking systems, and environmental upgrades. For many recyclers, especially smaller operators, these costs could outweigh the short- to medium-term returns, leading to financial strain or market exit. Without guaranteed demand for recycled materials or financial support for infrastructure upgrades, recyclers face a high-risk investment scenario that could slow down circular economy progress rather than accelerate it.

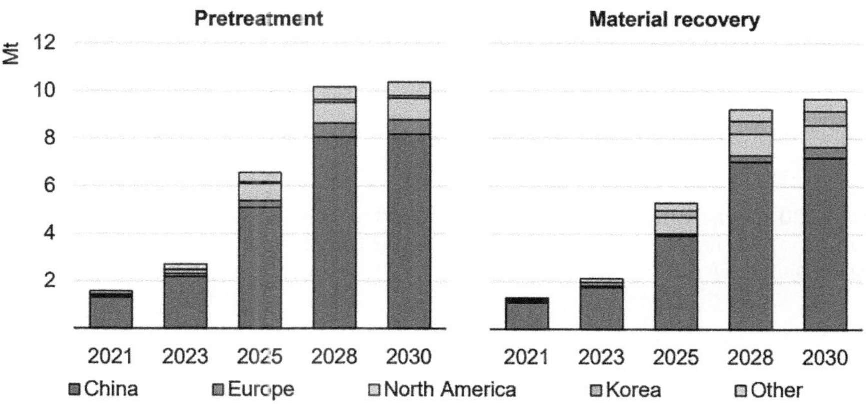

Figure 3.18 Battery recycling capacity based on announced projects, 2021–2030. *Source:* EV Battery Supply Chain Sustainability / IEA / CC BY 4.0.

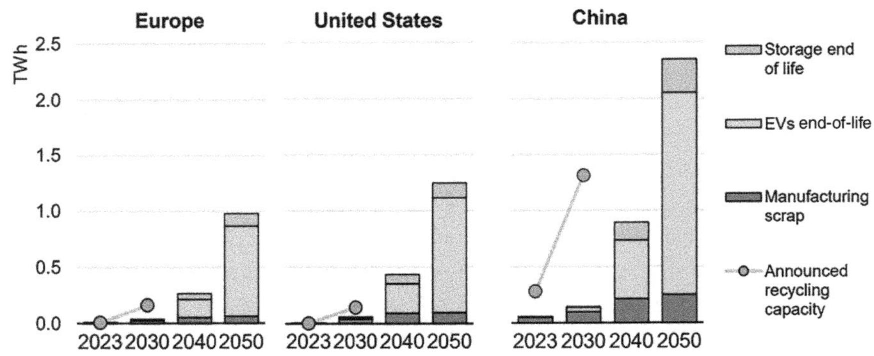

Figure 3.19 Maximum available battery recycling feedstock and recycling capacity by region in the Announced Pledges Scenario, 2023–2050. *Source:* EV Battery Supply Chain Sustainability / IEA / CC BY 4.0.

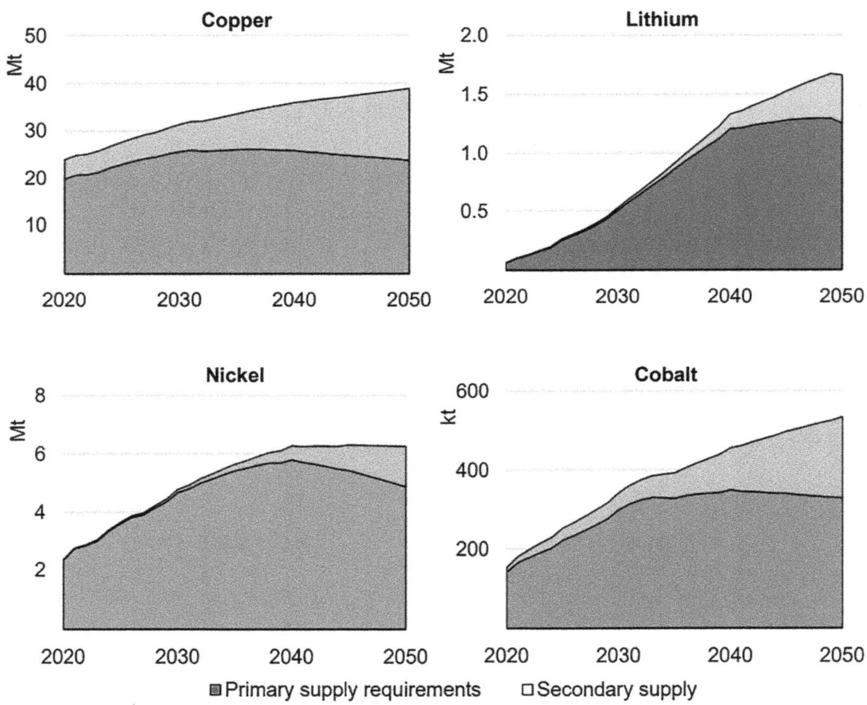

Figure 3.20 Primary supply requirements of key battery minerals with and without recycling in the Announced Pledges Scenario, 2020–2050. *Source:* EV Battery Supply Chain Sustainability / IEA / CC BY 4.0.

3.5.5.2 Regulatory and Compliance Risks

The Battery Passport system mandates greater transparency in material recovery, increasing the regulatory burden on recycling companies. Compliance requires accurate documentation of battery dismantling, material separation, and reintegration into the supply chain. Failing to meet these standards could result in

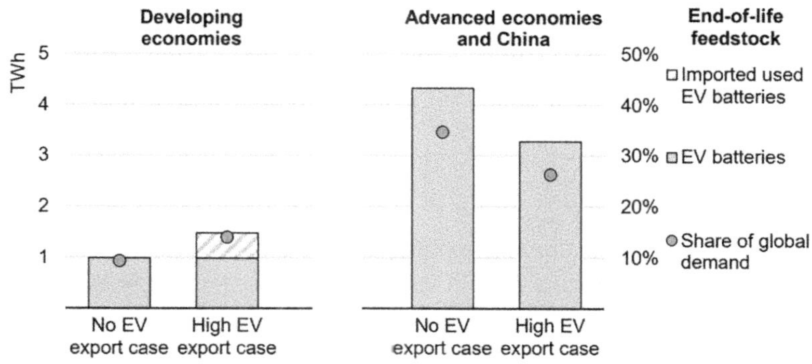

Figure 3.21 Potential impact of used electric vehicles export on end-of-life battery feedstock for recycling, 2050. *Source:* EV Battery Supply Chain Sustainability / IEA / CC BY 4.0.

financial penalties, legal action, or loss of business partnerships. Recycling firms operating in multiple regions must also navigate varying regulatory requirements, which could increase administrative costs and operational complexity. Inconsistent global regulations further complicate compliance efforts, particularly for companies exporting recycled materials across borders.

Battery Passport compliance introduces a new layer of regulatory complexity for recyclers. Detailed reporting on disassembly, recovery rates, and downstream material flow is not only time-consuming but also requires systems many recyclers currently lack. Failure to meet these requirements could trigger fines, reputational damage, or loss of certifications, especially in jurisdictions with strict oversight. For companies operating across multiple markets, the absence of uniform global regulations creates legal uncertainty, increasing compliance costs and the risk of operational misalignment.

3.5.5.3 Market Uncertainty and Demand Risks

The economic viability of battery recycling depends on demand for recovered materials such as lithium, cobalt, and nickel. Fluctuations in raw material prices pose a significant risk, as a decline in virgin material costs may reduce demand for recycled alternatives, making it difficult for recycling companies to sustain profitability. While the Battery Passport aims to incentivize circular economy practices, manufacturers may still prioritize lower-cost primary materials unless mandated to incorporate recycled content. Additionally, evolving battery chemistries may impact the value of certain materials, requiring recyclers to continuously adapt their processes to remain competitive.

The business model of battery recycling depends on predictable demand and favorable economics for recovered materials. If virgin material prices drop, recyclers may struggle to find buyers at competitive rates. Government regulations requiring manufacturers to use recycled content aim to keep demand relatively stable, but these regulations are subject to change. Implementation of these regulations may be postponed if there is an inadequate supply of recycled materials, increasing the risk for early adopters. Furthermore, innovations in battery chemistry may reduce reliance on high-value materials like cobalt. While the market remains unstable, recycling companies face volatile revenue streams and limited ROI on their process innovations.

3.5.5.4 Supply Chain Challenges

A reliable supply of end-of-life batteries is essential for recycling operations to scale effectively, yet current collection rates remain inconsistent due to low consumer participation, inefficient take-back systems, and logistical bottlenecks. These challenges create an unpredictable feedstock landscape, further compounded by the presence of informal or unregulated recyclers who divert valuable materials away from compliant operators. Operating with lower costs and fewer environmental safeguards, such informal actors undermine the competitiveness of formal recyclers, leaving them exposed to underutilized capacity, higher acquisition costs, and diminished returns on compliance investments. Without stable input flows, long-term business planning and profitability in the recycling sector remain uncertain.

3.5.5.5 Mitigation Strategies

Recycling companies can mitigate these risks by adopting cost-efficient processing technologies, leveraging partnerships with manufacturers to secure long-term supply agreements, and advocating for regulatory support to ensure a level playing field. Governments can play a crucial role in stabilizing the market through incentives, subsidies, or policy mandates requiring a minimum percentage of recycled content in new batteries. Improved consumer awareness campaigns and optimized collection networks can help increase the volume of end-of-life batteries entering the formal recycling stream. Addressing these challenges will be essential to ensuring that recycling companies remain viable and that the Battery Passport system successfully supports a circular battery economy (Figure 3.22, Table 3.8).

Table 3.8 Key economic risks associated with Battery Passport adoption by stakeholder group.

Stakeholder	Key economic risks
Manufacturers	High costs from new tech integration, compliance complexity, supply chain liability, and potential loss of price competitiveness.
Consumers	Increased product prices, data privacy concerns, cybersecurity vulnerabilities, and distrust in supply chain claims.
Governments	Heavy regulatory and infrastructure costs, uneven adoption risks, market disruption, and challenges enforcing compliance.
Recycling companies	Rising costs for upgrades and compliance, volatile demand for recovered materials, and no guarantee that the supply of end-of-life batteries will become stable.

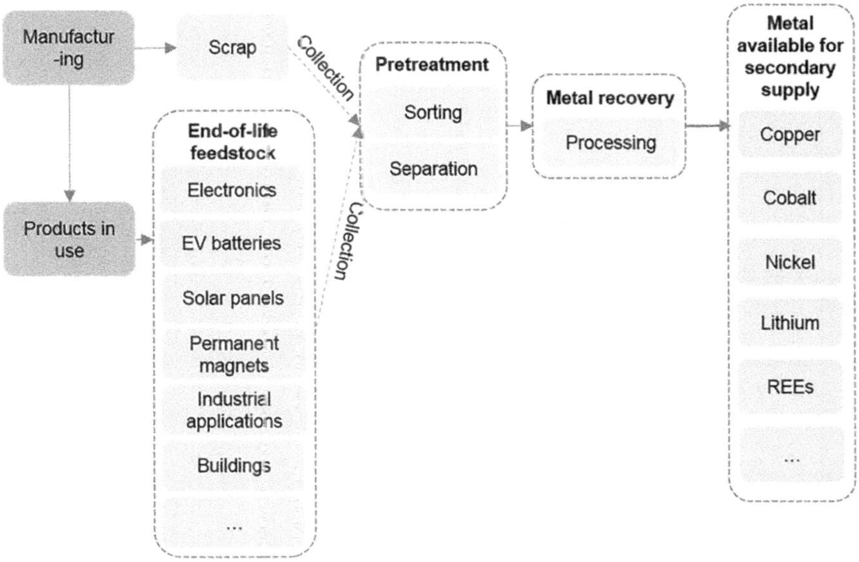

Figure 3.22 Overview of the recycling process for energy transition minerals. *Note:* REE, Rare earth elements. *Source:* Recycling of Critical Minerals IEA / IEA / CC BY 4.0.

3.5.6 Cost–Benefit Analysis for Stakeholders

3.5.6.1 Manufacturers

Benefits: The Battery Passport system allows manufacturers to track battery performance, optimize production processes, and ensure compliance with environmental regulations. By providing detailed data on battery materials and performance throughout the product lifecycle, manufacturers can improve resource efficiency and reduce waste, leading to cost reductions in manufacturing. Additionally, ensuring compliance with stringent sustainability requirements can enhance brand reputation and access to environmentally conscious markets [35].

Costs: Implementing the Battery Passport system involves significant initial investments in data management technologies, digital tracking tools, and compliance with transparency requirements. The costs associated with integrating new technologies and modifying existing manufacturing processes may lead to higher operational expenses [38].

3.5.6.2 Consumers

Benefits: Consumers benefit from the transparency provided by the Battery Passport, which offers reliable information on battery sustainability, performance, and composition. This allows consumers to make informed purchasing decisions, increasing trust in products with verified environmental credentials. Access to accurate lifecycle data also promotes the use of batteries with higher quality and longer lifespan [39].

Table 3.9 Cost–benefit risk analysis summary table.

Stakeholder	Costs	Benefits	Risks
Manufacturers	• **Initial Investment:** High costs for integrating data infrastructure, compliance technologies, and digital tracking tools. • **Operational Expenses:** Increased costs due to modifications in manufacturing processes to meet transparency and sustainability standards. • **Workforce Training:** Costs associated with hiring and training skilled personnel to manage new systems.	• **Market Access:** Enhanced access to markets that prioritize sustainable and transparent products. • **Resource Efficiency:** Improved resource management and reduced waste through optimized lifecycle tracking. • **Brand Reputation:** Strengthened brand image and consumer trust through compliance with environmental regulations.	• **Regulatory Compliance:** Potential penalties and reputational damage from noncompliance with evolving regulations. • **Market Resistance:** Risk of consumer pushback due to higher product prices. • **Supply Chain Risks:** Accountability for unethical or noncompliant practices within the supply chain.
Consumers	• **Higher Product Prices:** Increased costs for battery-enabled products due to compliance and technology integration costs passed on by manufacturers. • **Data Privacy Concerns:** Potential risks associated with the sharing and security of personal and usage data.	• **Informed Decisions:** Access to detailed information on battery sustainability and performance, enabling better purchasing choices. • **Product Quality:** Higher quality and longer lifespan of batteries due to improved manufacturing and recycling standards.	• **Cost Sensitivity:** Reduced adoption in price-sensitive markets due to higher upfront costs. • **Data Security:** Risks of data breaches and unauthorized access to personal information.
Recycling companies	• **Technology Upgrades:** Investment in advanced sorting and processing technologies to comply with Battery Passport requirements. • **Operational Adjustments:** Costs associated with adapting processes to handle detailed data on battery composition and lifecycle.	• **Material Recovery:** Increased recovery rates of valuable materials, enhancing profitability and sustainability. • **Process Optimization:** Improved efficiency in recycling processes through better traceability and data management.	• **Regulatory Burden:** Compliance with varying and evolving regulations across different regions. • **Market Demand:** Fluctuations in demand for recycled materials impacting profitability.
Government bodies	• **Regulatory Development:** Costs of creating and maintaining legal frameworks and compliance standards. • **Infrastructure Investment:** Public spending on digital and physical infrastructure to support the Battery Passport system.	• **Policy Enforcement:** Enhanced ability to monitor and enforce sustainability and transparency standards. • **Economic Growth:** Stimulation of job creation and economic activity in sustainable sectors.	• **Implementation Inequality:** Disparities in adoption and compliance across different regions. • **Budget Constraints:** Financial limitations in supporting infrastructure and regulatory development.

Costs: The increased production costs incurred by manufacturers to comply with the Battery Passport requirements may be passed on to consumers, leading to slightly higher prices for batteries and related products. While consumers may face higher upfront costs, the long-term savings from using durable and efficient batteries may offset these expenses [40].

3.5.6.3 Recycling Companies

Benefits: The Battery Passport provides detailed information on battery composition, simplifying the recycling process and potentially increasing the recovery rate of valuable materials such as lithium, cobalt, and nickel. By improving the traceability of materials, recyclers can optimize processes, reduce the costs of material separation, and enhance resource recovery rates [41].

Costs: To take full advantage of the Battery Passport system, recycling companies may need to invest in new technologies, infrastructure, and processes tailored to handling the specific data provided by the passports. These investments could pose challenges, particularly for smaller companies or those operating in regions with less regulatory support for recycling [42].

The Battery Passport system promises a transformative impact on the battery industry, offering substantial long-term benefits for manufacturers, consumers, and recycling companies. While the initial implementation costs are significant, the potential for cost reductions, resource optimization, and improved compliance with environmental standards creates a strong economic case for its adoption. Manufacturers can expect efficiency gains and access to more sustainable markets, while consumers will benefit from increased transparency and access to higher-quality products. Recycling companies, on the other hand, will see improvements in material recovery rates and cost efficiencies. Beyond these direct benefits, the Battery Passport system is likely to stimulate job creation in key sectors, contributing to the development of a green economy and driving innovation in sustainable practices across industries (Table 3.9).

3.6 Strategic Recommendations for Economic Adaptation

As the Battery Passport system gains momentum, stakeholders must navigate a complex landscape involving regulatory shifts, market dynamics, and evolving consumer behavior. To ensure economic viability and capitalize on emerging opportunities, strategic adaptation is crucial. The following recommendations for governments, businesses, and consumers are designed to help adapt to the changing economic trends and risks associated with the Battery Passport system.

3.6.1 Strategic Recommendations for Governments

To support the successful implementation of the Battery Passport, governments can play a significant role in creating a favorable regulatory environment, providing financial incentives, and facilitating infrastructure development. Most of those measures are already in place but may not be perfectly enforced in the future. We must determine which ones are key for the success of these regulations. Implementing those recommendations will help ensure a smooth transition for the Battery Passport.

3.6.1.1 Promoting Policy Coordination

Coordinating policies across regions is essential for reducing trade barriers and ensuring consistent implementation of Battery Passports. International cooperation can help avoid regulatory fragmentation, which can complicate compliance for businesses operating in multiple regions.

Institutions such as the World Trade Organization (WTO) and the International Electrotechnical Commission (IEC) can facilitate policy alignment by setting global standards for battery traceability and sustainability. This approach would promote regulatory consistency, reducing compliance costs for manufacturers and promoting smoother international cooperation.

Governments should engage in multilateral dialogues to align Battery Passport regulations and collaborate on global initiatives. Establishing regulatory alignment with major regions like the EU and China will help minimize trade disruptions and create a more predictable regulatory landscape for businesses.

3.6.1.2 Leveraging Financial Incentives

Financial incentives are key to encouraging companies, especially SMEs, to adopt the Battery Passport system. These incentives can offset the initial costs associated with compliance and encourage investment in sustainable technologies.

The EU's Green Deal provides financial support for industries that adopt sustainable practices, including Battery Passport compliance. By offering tax credits, grants, and subsidies, similar programs in other regions could accelerate the adoption of the Battery Passport system and facilitate the transition to a circular economy [43].

Governments should design financial incentives targeted at battery manufacturers, recyclers, and technology developers to promote the adoption of traceability technologies. Incentives could include subsidies for digital infrastructure development, tax breaks for companies meeting sustainability criteria, and grants for innovation in battery recycling.

3.6.1.3 Supporting Infrastructure Development

Governments should play a proactive role in facilitating infrastructure development through PPPs and direct investment.

Digital infrastructure is critical for tracking battery data throughout its lifecycle. Investments in technologies such as blockchain, IoT, and data analytics will enable real-time monitoring of battery conditions, supporting compliance with sustainability requirements and improving transparency [18].

They can also encourage the development of battery recycling facilities by providing subsidies or tax incentives to companies investing in advanced recycling technologies.

3.6.1.4 Establishing Enforcement Mechanisms

These entities must be empowered to conduct regular audits and impose a stronger ability to implement penalties on noncompliant actors. Smoothing the compliance mechanisms

Third-party Verification: Introduce independent auditors or certification bodies to verify the data recorded in Battery Passports, ensuring accuracy and adherence to established standards.

Penalties for Noncompliance: Implement a tiered system of penalties, including:

Fines: Monetary penalties for inaccurate or incomplete reporting of lifecycle data.

Tariffs: Increased import/export tariffs on batteries that fail to meet sustainability and traceability standards.

Trade Bans: Prohibit the sale or import of noncompliant batteries in regulated markets, creating significant economic disincentives for noncompliance.

3.6.1.5 Digital Monitoring Systems

Digital tools are vital for ensuring the enforcement of Battery Passport policies.

Blockchain Integration: Mandate the use of blockchain for creating immutable records of battery data, from material sourcing to end-of-life management.

AI-powered Monitoring: Deploy AI algorithms to identify discrepancies, flag high-risk actors, and optimize enforcement efforts.

Automated Compliance Alerts: Introduce systems that notify companies of potential noncompliance issues, reducing the risk of penalties while streamlining enforcement.

3.6.1.6 Enhancing Accountability Across the Supply Chain

To meet Battery Passport requirements, every link in the supply chain must comply with sustainability and transparency standards. This requires rigorous oversight of raw material sourcing, manufacturing, and recycling practices.

Mandatory Certification: Require suppliers to obtain certifications for sustainability, ethical sourcing, and recycling practices. Certification should be a prerequisite for participating in the Battery Passport system.

Incentives for Transparency: Offer preferential market access or reduced tariffs for batteries with verified, transparent supply chains.

Enforcement Through Trade Policies: Use trade policies to impose penalties on imports from suppliers that fail to meet sustainability and traceability standards.

3.6.1.7 International Collaboration

Global trade policies must incorporate Battery Passport compliance as a prerequisite for cross-border commerce. This ensures consistent enforcement and reduces regulatory fragmentation.

Mutual Recognition Agreements (MRAs): Develop agreements between countries to recognize and enforce each other's Battery Passport standards.

International Enforcement Bodies: Establish global organizations to monitor compliance and impose sanctions on noncomplying nations or companies.

Trade Restrictions: Introduce bans on batteries that fail to meet internationally recognized sustainability standards.

3.6.1.8 Leveraging Incentives for Enforcement

While penalties create deterrents, governments should also provide incentives to encourage compliance with Battery Passport requirements. These may include:

Subsidies for Compliance Costs: Offset costs associated with digital infrastructure, certification, and data management.

Tax Incentives: Reduce taxes for companies that demonstrate consistent compliance and superior sustainability performance.

Green Procurement Policies: Prioritize compliant companies in government contracts, providing an economic advantage to those meeting sustainability goals.

3.6.1.9 Penalties for Noncompliance

For stakeholders failing to meet Battery Passport requirements, governments can enforce significant consequences to drive accountability and compliance:

Fines: Impose steep monetary fines for inaccurate, incomplete, or fraudulent lifecycle data reporting.

Tariffs: Apply additional tariffs on noncompliant batteries, making them less competitive in the marketplace.

Trade Bans: Enforce strict prohibitions on importing, exporting, or selling batteries without verified Battery Passports.

Public Naming: Publish lists of noncompliant companies, creating reputational risks for brands that fail to adhere to sustainability standards.

3.6.1.10 Building Consumer Awareness and Market Pressure

Educated consumers are essential for reinforcing compliance with Battery Passport policies. Governments and NGOs should launch awareness campaigns to highlight the environmental and ethical benefits of purchasing Battery Passport-compliant products. Increased consumer demand for traceable, sustainable batteries will create market pressure on manufacturers to comply with these requirements.

By combining rigorous enforcement mechanisms with strategic incentives, governments can ensure the effective implementation of the Battery Passport system. Penalties such as fines, tariffs, and trade bans will deter noncompliance, while financial incentives and international collaboration will encourage broad adoption. These efforts will not only promote sustainability and transparency across the battery industry but also create a more equitable and responsible global marketplace.

To ensure the successful implementation of the Battery Passport system, governments should:

- Harmonize policies across regions to reduce trade barriers and compliance costs.
- Offer financial incentives like subsidies and tax breaks to ease adoption, especially for SMEs.

Table 3.10 Implementation measures.

Strategic area	Actions	Implementation measures
Policy coordination	Coordinate regulations internationally to minimize trade barriers and ensure consistent standards. Engage in multilateral dialogues.	Establish global standards via WTO and IEC; align with major regions (EU, China).
Financial incentives	Provide financial incentives targeting SMEs and key stakeholders to offset compliance costs and encourage sustainable investment.	Grants, subsidies, tax credits, and EU Green Deal-type initiatives.
Infrastructure development	Support the development of digital and physical infrastructure through investment and public–private partnerships.	Funding for blockchain, IoT, data analytics systems, and advanced battery recycling facilities.
Enforcement mechanisms	Establish specialized regulatory bodies to enforce compliance, conduct audits, and apply penalties. Introduce third-party verification systems.	Independent auditors; tiered penalties including fines, tariffs, and trade bans.
Digital monitoring systems	Deploy advanced technologies for real-time tracking and compliance enforcement.	Blockchain integration, AI-powered monitoring, and automated compliance alerts.
Supply chain accountability	Enhance transparency and accountability across the entire supply chain through mandatory certification and trade policy incentives.	Certifications for ethical sourcing, incentives like reduced tariffs, and penalties for nontransparent practices.
International collaboration	Foster global cooperation on Battery Passport compliance through mutual recognition agreements and international enforcement bodies.	MRAs, global compliance bodies, and international trade restrictions for noncompliant products.
Leveraging compliance incentives	Offer incentives that encourage compliance, offsetting compliance-related costs.	Subsidies, tax incentives, green procurement policies prioritize compliant companies.
Penalties for noncompliance	Impose clear and substantial penalties to deter noncompliance.	Monetary fines, increased tariffs, strict trade bans, and public naming and shaming campaigns.
Consumer awareness and market pressure	Educate consumers to create market demand for sustainable and traceable products, incentivizing industry-wide compliance.	Awareness campaigns led by governments and NGOs highlighting environmental and ethical benefits.

- Invest in infrastructure, including digital systems and recycling facilities, to support the transition.

By taking these steps, governments can drive a sustainable, competitive battery industry while minimizing economic disruption (Table 3.10).

3.6.2 Strategic Recommendations for Businesses

As the Battery Passport system gains momentum, businesses must strategically adapt to navigate regulatory changes, technological advancements, and shifting market expectations. The Battery Passport system will drive significant transformation across industries, and businesses that proactively embrace these changes will be well-positioned to benefit from emerging opportunities. This section provides actionable recommendations for businesses, focusing on key strategies such as building technological capabilities, engaging in cross-sector collaboration, and innovating new business models. By implementing these strategies, businesses can ensure compliance, optimize operations, and capitalize on the growing demand for sustainable products and practices within the battery value chain.

3.6.2.1 Building Technological Capabilities

To comply with the data collection and reporting requirements of the Battery Passport, businesses need to invest in advanced technologies. These technologies will ensure transparency, traceability, and compliance with emerging regulations.

3.6.2.1.1 Technological Integration and Strategy Companies should evaluate their current technological capabilities and develop a roadmap for integrating new hardware and software to support Battery Passport data management.

For example, Circulor, a company utilizing blockchain technology, tracks the provenance of raw materials in batteries, ensuring compliance with sustainability standards. This approach allows manufacturers to meet Battery Passport requirements while improving supply chain transparency [1].

3.6.2.2 Engaging in Cross-Sector Collaboration

Cross-sector collaboration is essential for maximizing the benefits of the Battery Passport. Businesses throughout the battery value chain, including manufacturers, recyclers, and technology companies, should work together to develop integrated solutions that align with global standards. Collaborative efforts can also help reduce costs and resource consumption while ensuring compliance with regulations.

3.6.2.2.1 Collaborative Framework Development Engaging in industry consortia such as the GBA can provide a platform for businesses to share resources, best practices, and data across the battery lifecycle. These collaborative efforts promote standardization and ensure the seamless exchange of information, thereby facilitating compliance with Battery Passport requirements [44].

Businesses should form stronger strategic partnerships with key stakeholders, including raw material suppliers and recycling facilities, to streamline the flow of information and achieve a circular economy. These partnerships can enhance supply chain transparency and reduce compliance costs through shared technology investments.

3.6.2.3 Innovating Business Models

The adoption of the Battery Passport system offers opportunities to explore new business models that capitalize on transparency and sustainability. Businesses can create additional value by leveraging Battery Passport data for novel revenue streams.

3.6.2.3.1 Exploring Subscription-based and Pay-per-use Models Companies can offer subscription-based services for battery tracking or battery swapping or pay-per-use models for accessing detailed data on battery performance and recycling rates to have a stronger grasp on their end-of-life systems. Data monetization opportunities, such as selling anonymized data for industry analysis, can also generate new revenue streams while promoting sustainability [45]. This is an example of new business models that can enable businesses to profit from this regulatory change.

The Battery Passport system signifies a transformative step in aligning the global battery industry with sustainability and circular economy principles. By enabling full lifecycle traceability, the system not only addresses pressing environmental challenges but also unlocks significant economic opportunities. Through enhanced resource efficiency, market transparency, and improved consumer trust, the Battery Passport has the potential to reshape the industry's competitive landscape, foster innovation, and drive sustainable growth.

However, its successful implementation depends on overcoming key challenges, including regional policy fragmentation, infrastructure gaps, and economic risks faced by manufacturers, consumers, and governments. The European Union's regulatory leadership provides a strong framework, but global adoption will require collaboration among regions, particularly with stakeholders in China, the United States, and developing nations. Harmonizing standards and fostering interoperability are essential to avoiding trade barriers and ensuring broad-based adoption.

Stakeholder engagement plays a critical role in this transition. Governments must create supportive policies, provide financial incentives, and invest in enabling infrastructure to promote compliance and adoption. Businesses need to innovate in technology and business models while also building cross-sector partnerships to share resources and reduce costs. Consumers, increasingly aware of sustainability issues, are poised to drive demand for certified products, reinforcing market incentives for transparency and accountability.

As global trade policies evolve and consumer behavior shifts, the Battery Passport system will serve as a critical tool for achieving economic viability and environmental sustainability. Its ability to harmonize trade practices, reduce material waste, and improve supply chain accountability positions it as a cornerstone of the global green economy.

To realize its full potential, stakeholders must adopt a collaborative and adaptive approach, addressing regional disparities, incentivizing adoption, and investing in technology that supports scalability and interoperability. By doing so, the Battery Passport system can become a universal standard, driving the battery industry toward a sustainable and equitable future, while serving as a blueprint for other sectors striving to balance economic growth with environmental stewardship.

References

1 Rufino Júnior, C.A., Sanseverino, E.R., Gallo, P., and Koch, D. (2024). Towards battery digital passport: reviewing regulations and standards for second-life batteries. *Batteries* 10 (4): 115. https://www.mdpi.com/2313-0105/10/4/115 (accessed 14 May 2025).

2 Rufino Júnior, C.A., Sanseverino, E.R., and Koch, D. (2024). Towards battery digital passport: reviewing regulations and standards for second-life batteries. *Batteries.* https://www.mdpi.com/2313-0105/10/4/115 (accessed 27 March 2025).

3 Berger, K., Schöggl, J.P., and Baumgartner, R.J. (2023). Digital battery passports to enable circular and sustainable value chains. *Journal of Cleaner Production.* https://www.sciencedirect.com/science/article/pii/S0959652622011131 (accessed 2 August 2025).

4 Malinauskaite, J., Anguilano, L., and Rivera, X.S. (2021). Circular waste management of electric vehicle batteries: legal and technical perspectives from the EU and the UK post-Brexit. *International Journal of Environmental Science and Technology.* https://www.sciencedirect.com/science/article/pii/S2666202721000161 (accessed 9 July 2025).

5 Nie, Y., Wang, L., Li, L., and Liao, H. (2023). Literature review on power battery echelon reuse and recycling from a circular economy perspective. *International Journal of Environmental Research and Public Health.* https://www.mdpi.com/1660-4601/20/5/4346 (accessed 18 April 2025).

6 Langley, D. J., Rosca, E., & Angelopoulos, M., Kamminga O., Hooijer C. (2023). Orchestrating a smart circular economy: guiding principles for digital product passports. *Journal of Business Research*, 169, 114259. https://www.sciencedirect.com/science/article/pii/S0148296323006185 (accessed 6 June 2025).

7 Harper, G.D.J., Kendrick, E., Anderson, P.A. et al. (2023). Roadmap for a sustainable circular economy in lithium-ion and future battery technologies. *Journal of Physics: Energy.* https://iopscience.iop.org/article/10.1088/2515-7655/acaa57/meta (accessed 21 August 2025).

8 Barkhausen, R., Fick, K., & Durand, A., Rohde C. (2023). Analysing policy change towards the circular economy at the example of EU battery legislation. *Renewable and Sustainable Energy Reviews*, 186, 113665. https://www.sciencedirect.com/science/article/pii/S1364032123005221 (accessed 30 May 2025).

9 da Silva, E.R., Lohmer, J., Rohla, M., and Angelis, J. (2023). Unleashing the circular economy in the electric vehicle battery supply chain: a case study on data sharing and blockchain potential. *Resources, Conservation and Recycling.*

https://www.sciencedirect.com/science/article/pii/S0921344923001052 (accessed 11 March 2025).

10 Bai, Y., Muralidharan, N., Sun, Y. K., & Passerini, S., Stanley Whittingham M., Belharouak I. (2020). Energy and environmental aspects in recycling lithium-ion batteries: concept of battery identity global passport. *Materials Today*, 41, 304, 315. https://www.sciencedirect.com/science/article/pii/S1369702120303060 (accessed 25 July 2025).

11 Fleischmann, J., Hanicke, M., and Horetsky, E. (2023). Battery 2030: resilient, sustainable, and circular. https://www.globalbattery.org/media/publications/battery-2030-resilient-sustainable-and-circular.pdf (accessed 30 May 2025).

12 Gebhardt, M., Beck, J., and Kopyto, M. (2022). Determining requirements for sustainable electric vehicle battery supply chain management: a mixed-methods approach. *Sustainable Production and Consumption*. https://www.sciencedirect.com/science/article/pii/S2352550922001701 (accessed 3 April 2025).

13 Berger, K., Baumgartner, R.J., Weinzerl, M., and Bachler, J. (2023). Data requirements and availabilities for a digital battery passport: a value chain actor perspective. *Journal of Cleaner Production* 4: 100032. https://www.sciencedirect.com/science/article/pii/S2666791623000052 (accessed 19 June 2025).

14 Kolhe, G. (2024). Sustainability and circularity metrics for EVs. https://lup.lub.lu.se. (accessed 8 August 2025).

15 Zhang, A., & Seuring, S. (2024). Digital product passport for sustainable and circular supply chain management: a structured review of use cases. *International Journal of Logistics Research and Applications*, 27, 2513, 2540. https://www.tandfonline.com/doi/abs/10.1080/13675567.2024.2374256 (accessed 16 May 2025).

16 Berg, H., Kulinna, R., and Stöcker, C. (2022). Overcoming information asymmetry in the plastics value chain with digital product passports. https://epub.wupperinst.org (accessed 28 June 2025).

17 Meyer, D., Schaupensteiner, N., and Riquel, J. (2024). Business models and ecosystems in the circular economy using the example of battery second use storage systems. www.mdpi.com (accessed 7 March 2025).

18 Ott, J., Schoeggl, J.P., and Baumgartner, R.J. (2024). End of life focused data model for a digital battery passport. www.sciencedirect.com (accessed 22 July 2025).

19 Naseri, F., Gil, S., Barbu, C., Cetkin E., Yarimca G., Jensen A.C., Larsen P.G., Gomes C. (2023). Digital twin of electric vehicle battery systems: comprehensive review of the use cases, requirements, and platforms. *Renewable and Sustainable Energy Reviews*, 179, 113280. https://www.sciencedirect.com/science/article/pii/S1364032123001363 (accessed 10 April 2025).

20 Tsolakis, N., Schumacher, R., & Dora, M., Kumar M. (2023). Artificial intelligence and blockchain implementation in supply chains: a pathway to sustainability and data monetisation? *Annals of Operations Research*, 327, 157, 210. https://link.springer.com/article/10.1007/s10479-022-04785-2 (accessed 31 August 2025).

21 da Silva, E.R., Lohmer, J., and Angelis, J. (2023). Unleashing the circular economy in the electric vehicle battery supply chain: a case study on data sharing and blockchain potential. *Resources, Conservation and Recycling.* https://www.sciencedirect.com/science/article/pii/S0921344923001052 (accessed 4 June 2025).

22 Yarali, A. (2021). Intelligent connectivity: AI, IoT, and 5G. https://books.google.com/books?hl=en&lr=&id=20pEEAAAQBAJ (accessed 26 May 2025).

23 Battery Pass Consortium (2024). *Press Release: First-of-its-Kind Analysis by Battery Pass Consortium Reveals Benefits – And Challenges – Of European Battery Passport [Internet].* Berlin: Battery Pass. https://thebatterypass.eu/news/press-release-first-of-its-kind-analysis-by-battery-pass-consortium-reveals-benefits-and-challenges-of-eu-passport (accessed 13 June 2025).

24 Berman, S.J. (2011). Not for free: revenue strategies for a new world. https://books.google.com/books?hl=en&lr=&id=-Ryh3Hs9pD0C (accessed 13 July 2025).

25 Henze, M. (2018). Cloud computing landscape. https://www.martinhenze.de/wp-content/papercite-data/pdf/hen18.pdf (accessed 12 May 2025).

26 Valotto, F. (2023). European Circular Economy Plan influence on real economy performances: a mixed methods study. https://unitesi.unive.it/retrieve/ef7cb948-630b-4db3-acb5-c9fcbfc222d9/888210-1274061.pdf (accessed 17 March 2025).

27 Küpper, D., Kuhlmann, K., Corey, A., and Saunders, M. (2022). Boosting resilience with production as a service. https://web-assets.bcg.com/7d/04/808072d148b9953576ffb020b20d/bcg-boosting-resilience-with-production-as-a-service-june-2022.pdf (accessed 24 August 2025).

28 Faktor, S. (2011). Econovation: the red, white, and blue pill for arousing innovation. https://books.google.com/books?hl=en&lr=&id=rok-Zr2ApQsC (accessed 1 April 2025).

29 Bellmann, C. (2021, December 9–10). The Circular Economy and International Trade: Opportunities for action in the World Trade Organization. 16th WCO PICARD Conference. Forum on Trade, Environment & the SDGs (TESS). https://na.eventscloud.com/file_uploads/b0085b981675b1efe7b72ddd0a2af6e1_ChristopheBellmannppt.pdf.

30 Dussaux, D., Dechezleprêtre, A., and Glachant, M. (2017). Intellectual property rights protection and the international transfer of low-carbon technologies. https://i3.cnrs.fr/wp-content/uploads/2017/12/WPi3-17-CER-05_rev.pdf (accessed 5 August 2025).

31 Melchor, A., and Shaparia, A. (2024). Trade-climate linkage in free trade agreements: a way forward for green growth. https://www.hks.harvard.edu/sites/default/files/centers/mrcbg/Final_AWP_237.pdf (accessed 26 May 2025).

32 Canadian Auto Dealer (2022). *Global Socio-Political Instability Creating Soaring Costs for EV Battery Materials [Internet].* Canadian Auto Dealer. https://canadianautodealer.ca/2022/11/global-socio-political-instability-creating-soaring-costs-for-ev-battery-materials/ (accessed 13 June 2025).

33 U.S. Energy Information Administration (EIA) (2025). *China Dominates Global Trade of Battery Minerals [Internet].* Today in Energy. https://www.eia.gov/todayinenergy/detail.php?id=65305 (accessed 13 June 2025).

34 Megyeri, E. and Tabajdi, G. (2024). Recollecting, reusing, and recycling: objectives and impacts of EU legislation on the battery value chain. *Business Logistics* *in Modern Economy.* https://www.researchgate.net/publication/384600285 (accessed 12 May 2025).

35 Berger, K., Schöggl, J.P., Baumgartner, R.J., and Bachler, J. (2023). Digital product passports to enable circular and sustainable value chains: conceptualization and use cases. *Journal of Cleaner Production* 353: 131492. https://www.sciencedirect.com/science/article/pii/S0959652622011131 (accessed 20 July 2025).

36 Uusitalo, T., Karhu, M., and Majaniemi, S. (2024). Digital product passports in circular economy: case battery passport. In: *EcoDesign for Sustainable Products, Services and Social Systems.* https://link.springer.com/chapter/10.1007/978-981-99-3818-6_13.

37 Orko, I., & Lavikka, R. (2024). Data platforms as tools for circular economy. In *Handbook for Sustainable Products, Services, and Social Systems* (ed. EcoDesign for Sustainable Products, Services and Social Systems). Springer Nature Singapore, 187-201. https://link.springer.com/chapter/10.1007/978-981-99-3818-6_14 (accessed 23 March 2025).

38 Schulz-Mönninghoff, M. and Evans, S. (2023). Key tasks for ensuring economic viability of circular projects: learnings from a real-world project on repurposing electric vehicle batteries. *Sustainable Production and Consumption* 35: 559–575. https://www.sciencedirect.com/science/article/pii/S2352550922003219 (accessed 9 April 2025).

39 Lieskoski, S., Tuuf, J., and Björklund-Sänkiaho, M. (2024). Techno-economic analysis of the business potential of second-life batteries. *Batteries* 10: 36. https://research.abo.fi/files/59422029/batteries-10-00036.pdf (accessed 22 April 2025).

40 Condemi, A., Cucchiella, F., and Schettini, D. (2019). Circular economy and E-waste: an opportunity from RFID TAGs. *Applied Sciences* 9: 3422. https://www.mdpi.com/2076-3417/9/16/3422/pdf (accessed 27 August 2025).

41 Koppelaar, R.H., Pamidi, S., and Hajósi, E. (2023). A digital product passport for critical raw materials reuse and recycling. *Sustainability* 15: 1405. https://www.mdpi.com/2071-1050/15/2/1405 (accessed 18 July 2025).

42 Jegham, J., and Sahlin Jankert, H. (2024). Exploring challenges and opportunities for circularity within the second life EV battery market. DiVA Portal. https://www.diva-portal.org/smash/get/diva2:1901157/FULLTEXT01.pdf (accessed 2 June 2025).

43 Terkes, M., Demirci, A., Gokalp, E., and Cali, U. (2024). Battery passport for second-life batteries: potential applications and challenges. https://ieeexplore.ieee.org (accessed 14 August 2025).

44 Langley, D.J., Rosca, E. and Angelopoulos, M. (2023). Orchestrating a smart circular economy: guiding principles for digital product passports. www.sciencedirect.com (accessed 6 April 2025).

45 Chirumalla, K., Kulkov, I., and Parida, V. (2024). Enabling battery circularity: unlocking circular business model archetypes and collaboration forms in the electric vehicle battery ecosystem. www.sciencedirect.com (accessed 30 May 2025).

4

Policy and Regulatory Framework: Enabling and Constraining the Battery Passport

4.1 Minimizing Compliance Costs and Enhancing Data Accuracy

The complexities of managing the battery lifecycle necessitate clear regulatory guidelines. By standardizing data collection and reporting requirements, frameworks such as the European Union's (EU) battery regulation ensure consistency across the battery value chain, including manufacturers, recyclers, and distributors. Standardization reduces the risk of misinterpretation, minimizes compliance costs for companies, and enhances the precision of environmental impact assessments [1].

4.2 Economic Benefits Through Coordination

Coordinated regulatory frameworks reduce trade barriers and provide companies with consistent practices across jurisdictions. This alignment supports international operations and ensures the global adoption of sustainable battery management practices. For instance, the standardization of Battery Passport protocols facilitates cross-border trade, making sustainability efforts recognized worldwide.

4.3 Fostering Stakeholder Trust and Transparency

Transparent and harmonized regulations build confidence among industry participants. Stakeholders are more likely to adopt sustainable practices when they trust the data shared through Battery Passports. Technologies such as blockchain can further enhance data security and integrity, ensuring reliable tracking of batteries throughout their lifecycle [2].

4.4 Supporting Circular Economy Initiatives

Regulatory frameworks, such as the EU Battery Regulation (2023/1542), play a crucial role in advancing the circular economy in the battery sector. These regulations enforce mandatory lifecycle documentation, extended producer responsibility (EPR), and recycling efficiency targets, ensuring that batteries are sustainably managed from production to end-of-life. By introducing digital product passports (DPPs), including the Battery Passport, these frameworks enhance data transparency, facilitate material traceability, and improve recycling processes. Additionally, minimum recycled content requirements for lithium, nickel, and cobalt, along with strict material recovery targets, promote waste reduction and resource reuse, aligning battery management practices with circular economy principles [3].

4.5 Reducing Regulatory Barriers to Trade

Internationally aligned standards, including efforts by the EU, the United States, and China, streamline compliance for companies operating in multiple markets. Standardization of technical and safety requirements reduces market uncertainties, facilitates the cross-border trade of second-life batteries, and enables more efficient reuse and recycling. Additionally, agreements such as the Basel Convention play a key role in defining the movement of hazardous materials, impacting the trade of batteries and their components. Regulatory alignment ensures that Battery Passports and related frameworks contribute to a globally integrated and sustainable battery supply chain.

4.6 Risks of Overregulation

As mentioned above with the economic risks of overregulation, we must note again that while robust regulatory frameworks are essential for guiding the battery industry toward sustainability and transparency, there is an increasing recognition – particularly in Europe – that overregulation can impose unintended burdens. Overly complex or rigid rules may hinder innovation, strain small- and medium-sized enterprises (SMEs), and create barriers that ultimately slow the transition to sustainable energy systems.

In the EU, stakeholders have begun to raise concerns that the layered and evolving compliance landscape, ranging from detailed Battery Passport requirements to strict recycling thresholds and EPR risks, overwhelms smaller actors who lack the financial and technical capacity to keep up. Unlike large multinationals with dedicated compliance teams, SMEs may find it difficult to invest in digital infrastructure, hire specialists, or adapt quickly to frequent regulatory updates. This could lead to market concentration, reducing competition and innovation across the sector.

Moreover, regulatory overreach may create friction in global trade. If European frameworks diverge too far from those in key markets such as the United States,

India, or Southeast Asia, interoperability and mutual recognition could be compromised. This fragmentation could complicate the global circulation of second-life batteries, materials, and battery management systems, slowing down the uptake of circular practices and undermining the very goals these regulations seek to advance.

There is also a risk that prescriptive compliance requirements, such as narrowly defined data formats, inflexible lifecycle reporting schedules, or mandatory use of specific technologies, may stifle experimentation and delay the adoption of novel solutions. Industry players increasingly call for a balance between regulatory precision and flexibility, allowing room for innovation while ensuring that core sustainability, safety, and transparency goals are met.

Striking the right balance between regulation and flexibility will be crucial to realizing the full potential of the Battery Passport. As Europe's experience shows, thoughtful calibration is needed to avoid inadvertently creating new obstacles on the path toward a sustainable and circular battery economy.

4.7 Global Fragmentation

Battery Passport policies vary across regions, shaped by local regulations, trade agreements, and sustainability goals. These regional variations influence how the Battery Passport system is implemented and operates globally. Understanding these differences is crucial for ensuring smooth adoption, especially in a highly globalized industry (Table 4.1).

4.7.1 China's Policy Vision

China is rapidly advancing its Battery Passport framework policies, driven by its focus on electrification and its position as the world's largest producer of electric vehicles (EVs). These policies emphasize traceability, recycling, and data transparency, aligning with China's domestic circular economy goals. However, as the leading supplier of batteries to the EU, China's battery manufacturers will need to adapt to the EU's stringent regulatory requirements, including the mandatory Battery Passport system set to be implemented by 2027. This interdependence between China's policies and EU regulations highlights the globalized nature of battery supply chains and the necessity for harmonized standards.

China is rapidly advancing its battery framework, driven by its focus on electrification and its position as the world's largest EV producer. The country's battery policy emphasizes traceability and compliance with environmental standards, aiming to support the circular economy and improve resource efficiency. China has launched pilot programs to digitize battery lifecycle management, emphasizing battery traceability, recycling, and data transparency in the EV supply chain. These objectives are aligned with those of the EU, making regulatory cooperation a viable possibility.

4.7.1.1 Emphasis on Battery Traceability and Recycling

China's policy framework prioritizes tracking batteries throughout their lifecycle, from production to end-of-life recycling. By enforcing mandatory disclosures on the

Table 4.1 Selected plans to phase out internal combustion engine (ICE) vehicles.

Region/country	Type	Description	Year
Canada	Policy	Auto manufacturers and importers must meet annual ZEV targets for LDV sales of 100% by 2035.	2023
Chile	Target	100% of sales of LDVs and public transport vehicles will be zero emissions by 2035, and by 2045 for M/HDVs.	2022
European Union	Policy	100% CO_2 emissions reduction for both new cars and vans by 2035.	2021
Israel	Target	Reduce average emissions from LDVs by 100% in 2050.	2022
Norway	Target	100% share of ZEVs in LDV sales by 2025 and in MDV sales by 2030.	2021
United Kingdom	Policy	100% of new car and van sales will be zero emissions by 2035.	2017
United States[4]	Policy	Advanced Clean Cars Rule II set a target for 100% ZEVs in LDVs by 2035.	2023
Automaker	Type	Description	Year
BMW	Target	50% of sales to be EVs before 2030.	2023
Changan	Target	60% of sales to be ZEVs by 2030.	2021
Ford	Target	50% of sales to be fully electric by 2030.	2022
Geely	Target	40% of sales to be electric by 2025.	2022
Honda	Target	100% of sales to be ZEVs by 2040.	2022
Hyundai, Genesis	Target	Annual EV sales of 2 million units and a BEV production share of 34% by 2030.	2023
Kia	Target	37% of sales to be EVs by 2030, reaching 1.6 million units per year.	2023
Mercedes-Benz	Target	50% of sales to be electric by 2030.	2024
Nissan	Target	55% of sales to be electric by 2030.	2023
Volkswagen	Target	50% of sales to be BEVs by 2030.	2023

Notes: ZEV, zero emissions vehicle; LDV, light-duty vehicle; MDV, medium-duty vehicle; M/HDV, medium-/heavy-duty vehicle; automaker, original equipment manufacturer of the automotive industry; BEV, battery electric vehicle. The year corresponds to the year of implementation. All automaker targets are for the global market.
Source: Batteries and Secure Energy Transitions (IEA) / IEA / CC BY 4.0.

sourcing and use of battery components, the framework ensures that manufacturers provide detailed information about materials, environmental compliance, and recycling rates. This traceability enables efficient resource recovery, reducing the environmental impact of battery disposal while improving recycling outcomes [4].

China has established the GB/T 34013-2017 and GB/T 34014-2017 standards, which provide strict criteria for battery dimensions, technical specifications, and coding systems. These standards facilitate lifecycle tracking by encoding essential data, such as manufacturer details, battery type, and production date. This approach not only ensures traceability but also streamlines sorting and recycling processes, making it easier to recover valuable materials from end-of-life batteries [5].

4.7.1.2 Government-Led Initiatives for Data Transparency

To improve transparency and accountability, the Chinese government has launched several programs to digitize the battery lifecycle. A key initiative includes the creation of a national database for tracking battery materials and performance. Additionally, the government is exploring blockchain technologies to enhance data accuracy and traceability, ensuring that the environmental impacts of battery production and use are comprehensively monitored.

In addition to its national database for battery tracking, the Chinese government has introduced the Administrative Measures for the Prevention of Pollution (MAPP), which enforces the principles of EPR. Under this framework, battery manufacturers are required to automate and standardize products to facilitate recycling, while recycling companies must disclose hazardous components, battery compositions, and environmental protection measures [5].

4.7.1.3 Pilot Programs for Digitizing Battery Lifecycle Management

China has implemented pilot programs to evaluate the effectiveness of digital Battery Passports in optimizing battery lifecycle management. These programs focus on digitizing data related to battery performance, durability, and recycling. By integrating this data into resource recovery systems, the pilot projects aim to increase recycling rates, minimize material waste, and reduce the carbon footprint of battery production.

China's comprehensive approach to battery lifecycle management provides a strong foundation for these pilot initiatives. Policies such as the GB/T 34013-2017 and GB/T 34014-2017 standards have, for example, the coding system standardizing data on battery manufacturer, type, production date, and other key metrics, facilitating efficient tracking and enabling precise sorting for reuse or recycling. The national battery management platform further supports these efforts by collecting and digitizing information on production, sale, use, and recycling.

Beyond its pilot programs, China has emphasized collaboration with organizations like the China Energy Storage Alliance (CNESA) and the China Electricity Council (CEC). These entities are instrumental in defining procedures for recycling and disassembly while promoting standards for integrating second-life batteries into renewable energy systems and low-power applications. These efforts enhance the potential for second-life batteries to contribute to the energy transition [5].

Pilot programs also align with China's principle of EPR, which mandates that EV and battery manufacturers establish service points for collecting and recycling used batteries. By leveraging digital platforms, these programs aim to enhance compliance with EPR policies and improve transparency in battery recycling systems. Furthermore, collaboration among government bodies, industrial alliances, and organizations like the CNESA highlights the potential of digital tools in integrating second-life batteries into renewable energy systems and other applications.

The lessons from these initiatives are expected to inform the nationwide rollout of digital Battery Passports equivalents, helping to standardize data sharing, improve material recovery, and align with China's sustainability and circular economy goals [5].

4.7.1.4 Aligning with Circular Economy Goals

China's battery policies are a critical component of its broader circular economy initiatives. By enforcing recycling standards and encouraging the development of advanced recovery technologies, the country promotes the reuse of valuable materials such as lithium, cobalt, and nickel. This approach not only minimizes environmental harm but also strengthens China's position as a global leader in sustainable battery management [6].

Incentives, such as government subsidies for energy-efficient vehicle purchases, complement these efforts by encouraging the adoption of new energy vehicles. On the other hand, punitive policies target manufacturers that fail to meet recycling or reuse targets, ensuring accountability throughout the battery lifecycle. This dual strategy of incentives and penalties reinforces China's circular economy objectives [5].

4.7.1.5 Future Directions for Policy Expansion

China's battery framework is expected to evolve as new policy priorities emerge, including second-life applications for batteries and stricter environmental targets. The integration of digital technologies, such as artificial intelligence and blockchain, will likely play a pivotal role in enabling more precise lifecycle tracking and resource optimization. These advancements will position Battery Passports as a cornerstone of China's long-term strategy for sustainable development and its commitment to global environmental goals.

Safety concerns remain a significant challenge as China addresses the use of second-life batteries. In response to safety incidents, the National Energy Administration has prohibited "large-scale" energy storage projects using second-life batteries. However, further clarification is needed to define what constitutes "large-scale," and future policies are expected to address these gaps to balance safety with sustainability goals [5].

China's Battery Passport-related regulations prioritize traceability and environmental protection, reflecting the country's broader focus on sustainable industrial development. Policies emphasize the safe disposal and recycling of batteries to prevent environmental contamination. For example, China's "Extended Producer Responsibility" scheme mandates that manufacturers remain accountable for the entire lifecycle of their products.

In addition to its strong domestic focus, China has invested heavily in developing battery recycling technologies, reducing reliance on imported raw materials while boosting its self-sufficiency in battery production. This strategy aligns with its long-term vision to dominate the global battery supply chain. As a result, China's approach combines environmental goals with economic and technological priorities, positioning the country as a major player in sustainable battery management [4].

China's efforts to implement a robust battery policy framework demonstrate its commitment to advancing sustainability and innovation in battery lifecycle management. Through initiatives focusing on traceability, recycling, and the digitization of battery data, China is making significant strides toward achieving its circular economy goals. However, its role as a primary supplier to the EU means that compliance with EU Battery Passport regulations will be critical. As the EU's regulatory framework sets the global benchmark for sustainability and traceability, China's alignment with these standards will not only secure its competitiveness in the European market but also contribute to the establishment of globally harmonized practices for sustainable battery management (Table 4.2).

4.7.2 North American Vision

North America does not currently have a formalized Battery Passport system. Instead, initiatives focused on battery traceability, recycling, and sustainability are emerging through public–private collaboration and alignment with existing trade agreements such as the United States-Mexico-Canada Agreement (USMCA). These efforts emphasize regional cooperation across the United States, Canada, and Mexico to promote environmental accountability and foster a circular economy.

Although the EU Battery Passport system is still under development, with implementation expected by 2027, it is already shaping global standards for battery lifecycle management. As a significant supplier of raw materials such as lithium, nickel, and cobalt for battery production, North America will also face indirect pressures to comply with EU guidelines. Batteries or raw materials originating in North America but destined for the EU market will need to meet the EU's strict traceability, carbon footprint, and sustainability requirements. This dynamic underscores the need for North American stakeholders to prepare for greater alignment with evolving international standards to remain competitive in the global market.

Historically, the United States has tended to follow the EU's lead on regulatory standards. For example, the adoption of General Data Protection Regulation (GDPR)-like privacy regulations in various U.S. states reflects a broader trend of aligning with EU frameworks to facilitate international business operations. Although the United States currently exports a limited number of batteries to the EU, batteries imported into both markets from China are often produced in the same factories. Manufacturers exporting to both the United States and the EU are likely to adapt all their batteries to the Battery Passport framework to streamline operations and logistics.

4.7.2.1 Transparency and Cross-Border Data Sharing

By enabling seamless tracking of battery components and recycling processes between the United States, Canada, and Mexico, these initiatives promote transparency across the entire battery value chain. This could become a new strategy for North America. This cross-border integration aligns with USMCA provisions on environmental cooperation, which prioritize sustainable practices and efficient resource management.

Table 4.2 Summary of the laws in China focusing on battery reuse and recycling.

Legislation	Year	Description
International transport of dangerous goods by road (in French, Accord européen relatif au transport international des marchandises Dangereuses par Route [ADR])	1957 (updated in 1975, 1985, 2011, and 2015).	ADR is an international agreement that aims to regulate the transport of dangerous goods to reduce the number of accidents caused by the transport of dangerous goods. This agreement defines dangerous goods' packaging, classification, labeling, and certification requirements. Dangerous goods are items and substances that are prohibited or that must be transported under specific conditions and are classified into nine classes, namely, toxicity, corrosivity, flammability, and reactivity. This agreement changed on 21 August 1975 and became effective on 19 April 1985. The agreement still underwent changes and updates in 2011 and 2015.
Law on the Prevention and Control of Environmental Pollution by Solid Wastes	1995 (updated in 2015 and effective in 2016).	This law aims to prevent and control pollution by solid waste and, for that purpose, prohibits the import, dumping, and disposal of solid waste except for cases where the government issues a license.
Law on Clean Production Promotion	2002 (updated in 2012)	This law forces companies to improve their processes to promote clean production. This law provides funds for lime production and can reduce or exempt small- and medium-sized companies from value-added taxes to promote clean production.
Waste Electrical and Electronic Equipment (WEEE) Pollution Prevention Administrative Measures (SEPA No. 40)	2008	These administrative measures aim to eliminate pollution and contamination relating to the disassembly, recycling, and disposal of electronic waste. These measures also present a scheme for licensing companies specializing in recycling electronic waste.
Interim Measures to Encourage the Purchase and Use of New Energy Vehicles in Shanghai	2014	The Shanghai municipal government provided financial incentives for recycling each EV battery. That incentive was 1000 RMB.
EV Battery Recycling Technology Policy	2015	Through this policy, the government aims to provide financial incentives for second-use companies and material extraction companies.
Circular Economy Promotion Law	2018	This law aims to promote the circular economy in companies, improve resource use efficiency, avoid environmental damage, and promote sustainable development. In order to achieve these objectives, this law implements a system of planning and statistical analysis of waste, and the current scenario implements fiscal and tax control and applies the EPR system. This law also provides tax incentives, monetary funds, loans, and credits to enable companies to implement the circular economy.
Pilot Scheme for EV Battery Recycling System in Shenzhen	2018	This public policy punishes industries that commit fraud or refuse to comply with recycling obligations. If the company does not fulfill its recycling obligations, its information will be inserted in the credit protection agencies.

Source: Rufino Júnior et al. [5] / MDPI / CC BY 4.0.

For instance, a digital infrastructure for tracking battery lifecycles ensures that key metrics such as material sourcing, performance, and recycling rates are accessible to stakeholders in all three countries. This level of transparency not only strengthens environmental accountability but also streamlines supply chain operations, fostering a more resilient and sustainable battery industry.

The United States has implemented the Universal Waste Law, which mandates proper collection, treatment, and recycling of lithium-ion batteries (LIBs) classified as hazardous waste. This law ensures that toxic components in LIBs are managed responsibly, avoiding environmental harm through improper disposal. Although this law governs battery management domestically, cross-border transparency initiatives, such as digital tracking systems, could align these efforts with regional sustainability goals under USMCA [5].

4.7.2.2 Alignment with USMCA's Environmental Provisions

The USMCA serves as a foundational framework for regional collaboration, providing guidelines to reduce emissions, improve environmental performance, and advance sustainable manufacturing. Battery Passport initiatives leverage these provisions to enhance lifecycle management, particularly within the rapidly growing EV industry.

By integrating Battery Passport policies with trade goals, North America can standardize sustainability measures, such as reporting carbon footprints and recycling efficiencies. This alignment helps bridge the gap between economic growth and environmental objectives, enabling the region to capitalize on the economic opportunities of the energy transition while adhering to its climate commitments under international agreements like the Paris Accord.

Recent federal initiatives in the United States, such as the establishment of the ReCell Center by the Department of Energy (DOE), emphasize the importance of recycling LIBs to reduce reliance on imported raw materials. The center focuses on developing advanced recycling technologies, improving the design of batteries for easier recycling, and integrating computational tools to validate recycled batteries' performance. These efforts directly support the USMCA's goals of enhancing environmental performance and fostering circular economy practices [5].

4.7.2.3 Collaborative Efforts Between Government and Industry

North America's Battery Passport initiatives stand out for their strong emphasis on public–private collaboration. Organizations such as industry leaders are partnering with policymakers to develop unified standards for battery lifecycle data reporting. These efforts are focused on improving transparency in areas such as:

Battery Durability and Performance: Ensuring long-term usability and minimizing waste.

Recycling Rates and Material Recovery: Enhancing circular economy practices by maximizing the recovery of valuable resources, such as lithium and cobalt.

Carbon Footprint Tracking: Providing detailed insights into emissions associated with battery production and lifecycle stages.

This collaboration fosters innovation by aligning government regulations with industry best practices, encouraging investment in sustainable technologies while building trust among stakeholders [7].

State-led initiatives also play a significant role in North America's battery lifecycle management. For example, California's Electronic Waste Recycling Act (EWRA) imposes an Advanced Recycling Fee (ARF) at the retail level for specific electronic devices. This fee funds recycling programs, ensuring that costs for handling and disposing of eligible devices, including batteries, are internalized. Additionally, the California State Assembly's House Bill No. 2832 establishes a producer responsibility principle, mandating that vehicle manufacturers cover the transportation and recycling costs of end-of-life EV batteries. Such measures encourage collaboration between government agencies and private industries to ensure sustainable battery management practices (Figure 4.1) [5].

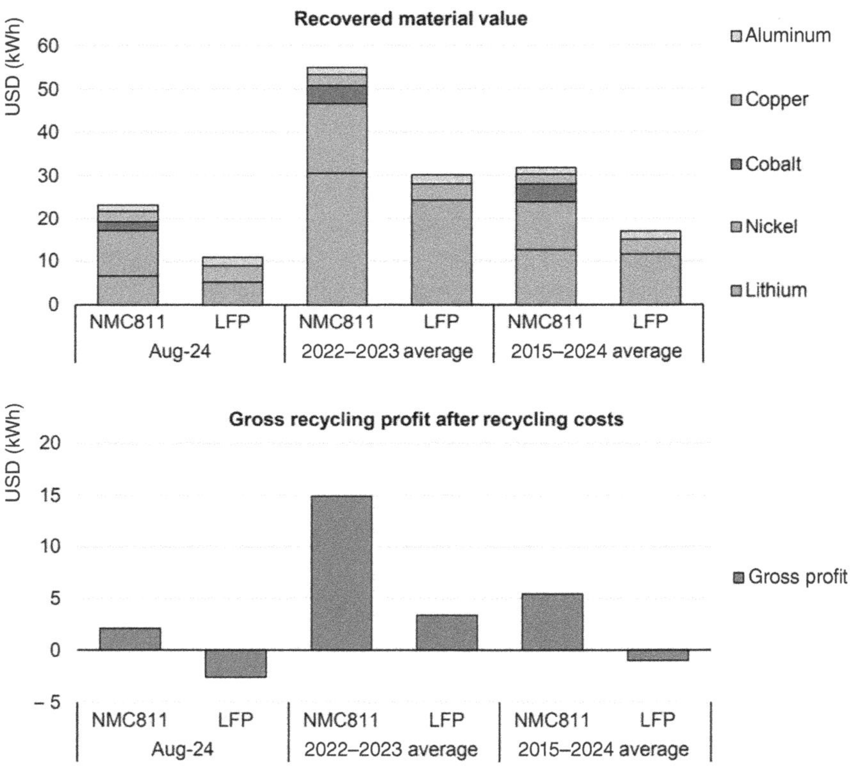

Figure 4.1 Indicative gross profit of battery recycling in the United States under different material price cases. Notes: Analysis using lithium carbonate, lithium hydroxide, nickel, cobalt, and copper prices from August 2024; the average prices from 2022 to 2023; and the average long-term prices from 2015 until August 2024. Based on the hydrometallurgy process, lithium hydroxide prices are used for NMC811 and lithium carbonate prices for lithium iron phosphate (LFP) batteries. Based on a market-based recycling business model, average domestic material recovery facility capacity in 2023 used 10 kilotons (kt) of battery packs for the United States. Includes black mass purchase cost for each chemistry. *Source:* Recycling of Critical Minerals IEA / IEA / CC BY 4.0.

4.7.2.4 A Distinctive Regional Approach

North America's approach to Battery Passports differs from other regions, such as the EU and China, by emphasizing flexibility and innovation over strict mandates. While the EU and China rely on centralized and mandatory regulatory frameworks, North America encourages voluntary standards and market-driven solutions. This strategy allows for greater adaptability, accommodating the diverse technological and industrial landscapes across the United States, Canada, and Mexico.

However, this decentralized approach also poses challenges, particularly regarding policy coordination. Differences in regulatory priorities between states, provinces, and countries can create inconsistencies that hinder seamless implementation. Addressing these disparities will require stronger coordination across federal and regional levels, along with a shared commitment to integrating sustainability into the battery value chain.

While North America's decentralized approach allows flexibility, it also highlights gaps in uniform policies. For instance, the lack of a federal regulation in the United States governing second-life applications for batteries has created challenges, as utility and state-level policies vary widely. Federal initiatives, such as the Federal Energy Regulatory Commission's (FERC) Order FERC 784 and Order FERC 792, have encouraged grid energy storage by classifying it as a generating resource. However, broader federal-level guidelines are still needed to standardize second-life applications and align them with regional and international frameworks [5].

In North America, Battery Passport-related policies remain fragmented and less centralized compared to the EU and China. The regulatory landscape is shaped by a mix of state- or province-specific initiatives and federal programs, particularly in the United States and Canada. For example, California has implemented stringent recycling requirements, while federal programs like the U.S. Department of Energy's "Battery Recycling Prize" focus on fostering innovation.

Rather than imposing strict mandates, North America emphasizes voluntary industry standards and market-based incentives to drive adoption of sustainable practices. While this approach encourages flexibility and innovation, the lack of a cohesive framework can create challenges for companies operating across multiple jurisdictions. As a result, North America's focus on economic incentives reflects a commitment to fostering private-sector leadership while gradually aligning with global sustainability goals [8].

4.7.2.5 Future Directions and Opportunities

Looking ahead, North America's Battery Passport initiatives are likely to evolve, driven by advancements in digital technologies such as **blockchain** and the **Internet of Things (IoT)**. These tools can enhance data accuracy and security, enabling real-time tracking of battery materials and performance metrics. Additionally, the region is expected to focus on:

Second-life Applications: Developing standards for repurposing used EV batteries in energy storage systems.

Stricter Recycling Targets: Introducing more ambitious goals to align with global sustainability benchmarks.

Integration with Renewable Energy Policies: Ensuring that battery manufacturing supports broader clean energy transitions.

By addressing these areas, North America has the potential to emerge as a leader in sustainable battery innovation, leveraging its unique strengths in technology, trade, and collaboration.

While North America's Battery Passport initiatives remain decentralized and market-driven, they represent an important step toward fostering sustainability in the battery industry. By focusing on transparency, cross-border collaboration, and voluntary compliance, these initiatives align with the region's trade and climate goals. However, with the EU in the process of establishing a mandatory Battery Passport framework, the global regulatory landscape is shifting toward standardized lifecycle management and traceability.

As North America continues to refine its Battery Passport initiatives, new areas of focus are emerging. For instance:

Second-life Battery Applications: The California advisory group has recommended standards for repurposing used EV batteries for energy storage systems, alongside producer responsibility principles for managing batteries at the end of their first useful life.

Recycling Innovations: The DOE's ReCell Center aims to enable direct recycling of cathode materials, reducing expensive reprocessing steps and fostering the economic viability of recycling systems.

Sustainable Supply Chains: Reports highlight the risks posed by North America's reliance on imported raw materials for LIBs. Policies emphasizing domestic sourcing and investment in material recovery are seen as critical to ensuring resilience in the battery supply chain [5].

In addition to its position as a significant player in battery production, North America's role as a major supplier of raw materials for batteries destined for the EU market further increases the need for compliance with EU guidelines. Even in the absence of a formalized North American Battery Passport, adherence to EU standards for raw material sourcing, carbon footprint reporting, and sustainability metrics will become increasingly necessary. By aligning its practices with these emerging requirements, North America can maintain its competitiveness in international markets while fostering innovation and leadership in sustainable battery management (Table 4.3).

4.7.3 African Battery Initiatives

Africa is an emerging market for second-life battery applications, driven by its unique energy challenges. Many regions experience frequent blackouts and brownouts, making reliable energy storage solutions critical for economic development. Second-life batteries integrated with photovoltaic systems hold significant potential to improve energy access and reliability.

For instance, the Faraday Battery Challenge project in Kenya has implemented energy storage systems using second-life batteries to power homes with integrated

Table 4.3 Summary of the principal regulations in the area of batteries in different regions of the world.

Status	China	Europe	United States	Japan	Korea	Africa	South America	India
Regulations on EV batteries for recycling	✓	The Battery Directive 2008/98/EC allowed batteries to be transferred from one party to another without being declared waste.	—	—	Clean Air Conservation was announced in 2004 and implemented in 2005 to monitor the pollution of cars with diesel emissions.	—	—	—
General regulation on LIB batteries	✓	✓	The United States has federal waste regulations that regulate batteries according to the status of each type of battery.	The Law on the Promotion of the Effective Use of Life Resources promotes the reduction of waste, reuse, and recycling of batteries.	—	—	—	—
Regulations on second-life batteries	✓	The Battery Directive 2008/98/EC allowed batteries to be transferred from one party to another without being declared waste.	—	—	Clean Air Conservation was announced in 2004 and implemented in 2005 to monitor the pollution of cars with diesel emissions.	—	—	—

(Continued)

Table 4.3 (Continued)

Status	China	Europe	United States	Japan	Korea	Africa	South America	India
EPR	✓	✓	✓	—	—	—	Some countries, such as Brazil, but not all countries in South America	—
Recycling efficiency target	✓	—	—	—	—	—	—	—
Leading players who are investing in second-life battery projects	The legislation establishes that the efficiency for recycling Ni, Co, and Mn must be 96% for the hydrometallurgical process and 97% for the pyrometallurgical recycling process to obtain nickel and rare earth.	The battery directive states that 50% of the total weight of batteries that reach the end of their useful life must be recycled.	—	—	—	—	—	—

Leading players who are investing in second-life battery projects	Yinlong Energy, Build Your Dream (BYD), GreatWall Power	Daimler GETEC/The Mobility House Remondis/ EnBW (Germany), Renault (UK), Umicore (UK), Connected Energy (UK), Relectrify (Australia), Bosch (Alemanha), Siemens (Alemanha), Vattenfall (Germany), BMW (Germany), Audi (Germany), Volkswagen (Germany), Fortum (Finlândia), Acceleron (UK)	General Motors, ABB, Spires New Technologies Inc. (SNT), Chevrolet, Florida Power and Light, and FreeWire	4R Energy, Honda, IT Asset Partners (ITAP), Mitsubishi, and Nissan	—	Eaton	—	—
Major recycle associations of LIB	Waste Battery Recycling Committee China Battery Industry Association	European Battery Recycling Association and Recharge	NAATBatt	Battery Association of Japan	Korea Battery Industry Association	The South African EV Association is still looking for viable projects for battery recycling.	—	—

Source: Rufino Júnior et al. [5] / MDPI / CC BY 4.0.

solar energy. These projects highlight the potential of second-life batteries to address energy reliability issues and support sustainable development. However, Africa currently lacks a comprehensive regulatory framework for end-of-life battery management. Without adequate legislation and infrastructure for collection and recycling, the risk of improper battery disposal persists. Developing these systems could unlock immense opportunities for second-life battery adoption in the region [5].

4.7.3.1 International Collaboration

Achieving true harmonization requires aligning regional policies through international collaboration. Trade agreements, such as the Comprehensive and Progressive Agreement for Trans-Pacific Partnership (CPTPP), and organizations like the United Nations Economic Commission for Europe (UNECE), play a pivotal role in fostering cross-border cooperation. The EU's Battery Passport framework, for example, could serve as a blueprint for other regions, encouraging mutual recognition of standards and creating opportunities for joint development of digital traceability systems. Collaborative efforts under global initiatives, such as ISO standards for sustainable batteries, further support integration by establishing shared benchmarks for safety, durability, and environmental compliance. By leveraging these platforms, nations can address disparities and work toward a unified global framework.

4.7.3.2 Challenges to Standardization

Despite progress, significant barriers hinder the alignment of Battery Passport policies across regions. Variations in technological capabilities, such as the lack of advanced digital infrastructure in developing economies, pose challenges for implementing robust traceability systems. Additionally, regulatory priorities differ widely. While the EU emphasizes environmental goals, other regions may focus on economic growth or energy security, creating conflicts in setting universal standards. Economic constraints, especially for smaller economies, further complicate compliance with stringent regulations. Overcoming these challenges requires targeted support, such as technology transfer programs, financial assistance for infrastructure development, and active dialogue among policymakers to balance regional needs with global objectives.

4.7.3.3 Africa's Path Forward with Battery Passports

Africa's journey toward adopting Battery Passports is still in its early stages, but the continent's strategic importance as a producer of critical battery raw materials places it at the heart of the global push for sustainability and supply chain transparency. With vast reserves of cobalt, lithium, and other key materials, Africa's active participation in Battery Passport initiatives has the potential to transform mining practices, improve resource efficiency, and support local economic development. Collaborative efforts by organizations such as the African Development Bank (AfDB) are driving initiatives to align Battery Passport adoption with regional priorities like sustainable mining and e-waste management.

4.7.3.4 Africa's Growing Role in Battery Raw Material Production

Africa is home to some of the world's largest reserves of critical minerals required for battery production, making its role in the global supply chain indispensable. For example, the Democratic Republic of the Congo (DRC) supplies over 70% of the world's cobalt, while Zimbabwe has significant lithium reserves.

Adopting Battery Passports in Africa can improve transparency and accountability in the sourcing of these materials, ensuring compliance with international standards for responsible mining.

Enhancing transparency and accountability in the sourcing of critical minerals in Africa is essential for ensuring compliance with international standards for responsible mining. This includes addressing environmental impacts and labor practices, such as child labor and unsafe working conditions, particularly in artisanal mining. By implementing robust governance frameworks and traceability mechanisms, African nations can attract investment from companies that prioritize ethically sourced materials, thereby fostering a more sustainable and responsible mining sector [9].

4.7.3.5 Initiatives for Sustainable Mining and Electronic Waste Management

The shift toward sustainability is gaining momentum in Africa's mining sector, driven by both local and international pressures to reduce the environmental and social costs of extractive industries. Battery Passports provide a practical tool for tracking materials from extraction through to disposal, enhancing accountability and efficiency in material use.

The challenges of electronic waste (e-waste) management in African nations are significant, with large volumes of discarded batteries and electronics contributing to environmental hazards. Integrating initiatives like Battery Passports with e-waste management systems can help in identifying and recovering valuable materials, thereby reducing the ecological impact of disposal practices [10].

4.7.3.6 Efforts to Integrate Battery Passports with Local Sustainability Goals

Africa's unique energy challenges, particularly in regions experiencing unreliable electricity supply, present opportunities for leveraging second-life battery applications. Projects such as the Faraday Battery Challenge have highlighted the potential of repurposing used EV batteries for energy storage. In Kenya, for instance, second-life batteries have been successfully paired with solar photovoltaic systems to provide reliable power in off-grid communities.

Battery Passports can support these initiatives by offering precise data on battery performance, remaining capacity, and safety, ensuring second-life applications align with both sustainability and economic development goals. Such projects demonstrate the dual benefits of improving energy access while extending the lifecycle of battery materials [5].

4.7.3.7 Challenges and Opportunities in Implementation

While the adoption of Battery Passports offers transformative potential for Africa, several challenges must be addressed to enable successful implementation:

Infrastructure Development: Many African nations lack the advanced digital infrastructure required for implementing Battery Passport systems, such as blockchain-based traceability or lifecycle data monitoring. Investments in technology and capacity building will be essential to bridge this gap.

Policy Fragmentation: With over 50 countries across the continent, each with distinct regulations and enforcement capabilities, achieving policy alignment is a major hurdle. Regional bodies like the African Union (AU) and Southern African Development Community (SADC) will need to spearhead efforts to establish standardized regulations and enforcement mechanisms.

Financial Barriers: Developing nations may face funding constraints for adopting Battery Passport systems. Partnerships with international organizations, such as the AfDB and the Global Battery Alliance, could provide the financial and technical support needed to drive implementation.

Despite these challenges, the opportunities for Africa are immense. By embracing material passports, the continent can position itself as a global leader in sustainable resource management, attract ethical investments, and support innovation in green technologies.

4.7.3.8 Looking Ahead: Africa's Role in a Global Ecosystem

Africa's integration into the global Battery Passport system represents an opportunity to redefine its role in the battery supply chain, not just as a raw material supplier, but as a leader in sustainable and transparent practices. Moving forward, priorities should include:

Establishing regional centers of excellence for battery recycling and second-life applications. Collaborating with international stakeholders to develop capacity-building programs that enhance local expertise. Leveraging digital technologies like blockchain and IoT to modernize material tracking systems.

By addressing these priorities, Africa can unlock the full potential of Battery Passports, contributing not only to regional sustainability but also to the success of global circular economy initiatives (Table 4.4).

4.7.4 Latin American Initiatives

Latin American countries show strong potential for the adoption of EVs but continue to face major challenges in managing batteries at the end of their life cycle. Most countries in the region lack the necessary infrastructure for the collection, treatment, and recycling of EV batteries. This absence of appropriate systems increases the likelihood of improper disposal, posing environmental and public health risks [5].

Countries such as Brazil, Chile, Colombia, Costa Rica, Paraguay, and Peru have introduced regulations and host industries that recycle lead–acid batteries. However, their limited technical and operational capacity means they are often unable to process the growing volumes of lithium-ion and other advanced batteries. As a result, exporting end-of-life batteries for treatment abroad becomes necessary, raising significant legal and logistical concerns from a regulatory perspective [5].

Adding to this challenge is the illegal or irregular import of electronic waste, including used batteries disguised as second-hand goods for reuse. Inadequate regulation and weak customs enforcement enable large volumes of waste to enter

Table 4.4 Policies and measures for the recycling of critical minerals.

	Strategic plans	Financial incentives	EPR	Cross-border trade
Australia*	•	•	•	•
Brazil		•	•	
Canada*	•	•	•	•
Chile		•	•	•
China	•	•	•	•
Colombia			•	
European Union	•	•	•	•
France		•	•	
Germany	•	•	•	
Ghana			•	•
India	•		•	•
Indonesia				•
Italy	•	•	•	•
Japan		•	•	•
Kenya	•			
Korea	•		•	•
Nigeria			•	•
Pakistan				•
South Africa			•	•
United Kingdom	•	•		•
United States*		•	•	•
Viet Nam			•	•

* In countries with federal systems, mineral recycling policies are typically managed at the subnational level rather than the national level. In the table, a circle is placed if the Policy Tracker includes policies from states that have implemented a policy related to the category.
Source: Recycling of Critical Minerals IEA / IEA / CC BY 4.0.

Latin American markets under the false pretense of reusability. The lack of clear distinctions between genuinely reusable electronics and waste products undermines inspection efforts and renders EPR frameworks ineffective [5].

Nonetheless, there is growing interest in strengthening EPR systems, implementing e-waste collection programs, and developing practical solutions for battery reuse. In countries like Brazil, large land areas and favorable conditions for solar and wind energy create an opportunity to develop business models that integrate second-life batteries with renewable energy systems, offering both economic and environmental benefits [5].

Brazil's National Solid Waste Policy (NSWP), established under Law No. 12.305/2010, represents one of the most advanced legal instruments in the region. It introduces the principle of shared responsibility across producers, importers, distributors, retailers, and consumers for the entire product life cycle. The law mandates systems for the collection and return of used products, enabling their reintegration into production cycles or ensuring environmentally sound final disposal [5].

Under this legislation, commercial outlets and authorized technical support services that handle portable batteries, particularly lead–acid, nickel–cadmium, and mercury oxide types, are legally required to accept used batteries from consumers. Once collected, producers and importers are responsible for ensuring their proper recycling or disposal. The law also requires the implementation of a reverse logistics system, which must operate independently from public waste management services. This system aims to institutionalize accountability among all actors involved in the life cycle of battery products [5].

4.7.5 Asia-Pacific (APAC) Region Initiatives

South Korea is a global leader in battery production, with major manufacturers like LG Energy Solution, Samsung SDI, and SK ON driving innovation in energy storage technologies. The country has also been at the forefront of second-life battery pilot projects. For example, a program initiated in 2015 explores the use of second-life batteries for stationary energy storage and backup systems.

South Korea's regulatory landscape includes the EPR policy, implemented in 2002, which mandates recycling targets and imposes fines for noncompliance. However, used batteries are classified as waste under South Korean law, which increases transportation and recycling costs and limits second-life battery applications. Experts suggest that the responsibility for recycling and reuse should shift from the government to private companies to enhance efficiency and innovation in battery lifecycle management [5].

Additionally, South Korea's Waste Management Act and Economy Promotion and Resource Recycling Law have contributed to advances in recycling practices. However, more comprehensive policies are needed to support second-life battery markets and address the logistical challenges of waste classification. These advancements would enable South Korea to better align its practices with global sustainability standards and maintain its leadership in battery innovation [5].

The Asia-Pacific region faces unique challenges in harmonizing Battery Passport regulations due to its diverse economic landscape. Countries like Japan, South Korea, and Australia are leading initiatives to standardize battery recycling and environmental compliance. However, differences in regulatory enforcement across ASEAN countries could create challenges in achieving consistent Battery Passport adoption.

4.7.5.1 Harmonizing Regulations in Asia-Pacific

The Asia-Pacific region represents a complex yet vital area for Battery Passport adoption. As a hub for global battery production and consumption, this region faces unique challenges in aligning regulations due to its diverse economic landscape, varying levels of development, and fragmented regulatory frameworks. However, regional cooperation and trade agreements present promising opportunities to drive harmonization and advance sustainability goals.

4.8 Key Challenges

4.8.1 Fragmented Regulatory Environments

The Asia-Pacific region comprises a mix of advanced economies, such as Japan and South Korea, alongside developing nations in ASEAN. This diversity results in highly fragmented regulatory environments, where standards for battery recycling, lifecycle tracking, and environmental compliance differ significantly between countries.

For example, Japan and South Korea have established stringent recycling laws and advanced battery management systems, while some ASEAN countries still grapple with basic infrastructure for waste management. This lack of uniformity complicates efforts to implement a seamless Battery Passport system across borders, as manufacturers must navigate conflicting regulations that increase compliance costs and operational inefficiencies.

4.8.2 Lack of Coordination Between Developing and Developed Nations

A major barrier to harmonization lies in the disparity between developed and developing countries within the region, similarly to the rest of the world. Developed nations such as Australia and Singapore often adopt stricter environmental standards and advanced technologies, while developing economies may lack the resources, infrastructure, or institutional capacity to enforce similar regulations.

This imbalance creates a regulatory divide, where less stringent enforcement in some countries may undermine the effectiveness of regional initiatives. Without addressing this disparity, achieving consistent battery passport adoption across Asia-Pacific will remain a significant challenge [11].

4.8.3 Opportunities for Harmonization Through Regional Cooperation

4.8.3.1 Regional Forums and Initiatives

Efforts to align environmental policies and battery standards are underway through regional platforms such as the ASEAN Economic Community (AEC) and Asia-Pacific Economic Cooperation (APEC). These forums provide opportunities for member countries to collaborate on developing shared policies and best practices for battery lifecycle management.

For instance, APEC's initiatives on environmental sustainability could serve as a foundation for creating unified guidelines for Battery Passports, focusing on resource efficiency, material recovery, and lifecycle tracking. By leveraging these platforms, the region can bridge policy gaps and promote a coordinated approach to Battery Passport implementation.

4.8.3.2 Emphasis on Regional Trade Agreements to Facilitate Policy Alignment

Trade agreements such as the CPTPP play a pivotal role in promoting harmonization. These agreements establish common standards for product safety, environmental protection, and trade facilitation, making them valuable tools for aligning Battery Passport regulations.

For example, CPTPP's provisions on environmental cooperation could be expanded to include battery lifecycle management, encouraging member countries to adopt consistent practices. Such alignment would reduce trade barriers, streamline compliance, and enhance regional supply chain efficiency, ensuring that battery-related policies support broader sustainability goals.

4.8.3.3 Promising Developments and Future Outlook

Despite the challenges, several Asia-Pacific countries are taking proactive steps to drive harmonization:

Japan and South Korea's Leadership: These nations are spearheading initiatives to integrate battery systems into their advanced recycling frameworks, showcasing the feasibility of lifecycle tracking technologies. Their leadership can provide a model for other countries in the region to follow.

ASEAN's Collaborative Efforts: ASEAN countries are working toward aligning their environmental policies through collective action plans under the UN's **3R (Reduce, Reuse, Recycle) framework**. Expanding these efforts to include Battery Passport standards could address key policy disparities and promote sustainable battery management practices.

Digital Technologies for Policy Integration: The adoption of blockchain, artificial intelligence, and IoT technologies is gaining traction in the region, offering innovative solutions for tracking battery lifecycles. These tools can enhance transparency and ensure that Battery Passports align with regional and global standards.

Harmonizing regulations in Asia-Pacific is a complex but achievable goal, requiring a combination of regional cooperation, trade agreement alignment, and technology-driven solutions. By leveraging forums like APEC and agreements such as CPTPP, the region can address policy fragmentation and promote the adoption of consistent Battery Passport standards. These efforts will not only advance sustainability objectives but also position Asia-Pacific as a key player in the global transition to a circular economy for batteries (Table 4.5).

What emerges clearly from this global analysis is the profound fragmentation that currently defines Battery Passport policies worldwide. While regional efforts – from China's state-driven digital traceability programs to the EU's mandatory regulatory frameworks and North America's voluntary, market-driven initiatives – demonstrate significant strides toward sustainability, they also reveal the lack of cohesion in implementation. Each region advances with its own priorities, shaped by domestic policy goals, technological capacity, economic structure, and political context.

This fragmentation presents a serious barrier to the development of a truly global and interoperable Battery Passport system. Manufacturers operating across borders must navigate conflicting regulations, varying data standards, and inconsistent enforcement mechanisms. This not only inflates compliance costs and creates market distortions but also undermines the core objective of the Battery Passport: to enable transparent, circular, and sustainable battery value chains on a global scale.

Table 4.5 Comparative analysis of second-life battery market regulations across different regions.

Aspect/region	United States	South Korea	Africa	Latin America	Europe	China	Similarities
Principle of producer responsibility	• Implemented at the state level with variations. • Focus on electronic waste management programs.	• Strongly implemented with DRS and EPR policy.	• Not widely implemented or in initial stages.	• Brazil stands out with the NSWP. • Other countries are developing or seeking to implement similar principles.	• Broad implementation with detailed regulations for batteries and electric vehicles. • Battery Directives establish producer responsibility for the collection, treatment, and recycling of batteries.	• Robust closed-loop management system from production to disposal.	• EPR as a basic principle in all regions.
Incentives and penalties	• Varies by state; incentives for energy storage, but without a unified approach to recycling penalties or incentives. • FERC orders recognize energy storage as a resource. • Incentives for energy storage connection to the grid.	• EPR establishes collection targets and fines for noncompliance, encouraging recycling. • Incentives for battery recycling.	• Lack of clear or implemented incentive and penalty structure.	• Inadequate infrastructure and a lack of clear regulation limit the effectiveness of potential incentives or penalties.	• Establishes clear collection and recycling efficiency objectives, with incentives for reuse and penalties for noncompliance. • Strict standards promote recycling and the circular economy. • Specific targets for recycling efficiencies.	• Incentives and mandates for battery manufacturers regarding recycling.	• Use of incentives and penalties for compliance.
Specific regulations for recycling and reuse	• Battery Law, state regulations like EWRA in California.	• Focus on reuse and recycling, but gaps for second-life batteries.	• N/A	• Variety of national regulations and some lead-acid battery recycling initiatives.	• N/A	• N/A	• Variety of approaches focusing on recycling.

(Continued)

Table 4.5 (Continued)

Aspect/region	United States	South Korea	Africa	Latin America	Europe	China	Similarities
Differences	• Regulatory fragmentation between states. • The "Battery Act" does not include LIBs and NiMH unless they contain heavy metals. • No specific incentives for second-life battery applications.	• Government responsibility for recycling, different from the more common producer responsibility model.	• N/A	• Regulatory diversity, with some countries advancing more than others. • Regulatory frameworks are less developed compared to Europe.	• Comprehensive and uniform approach across the EU. • Strong emphasis on circular economy and sustainability. • Regulations like the EU Battery Directive address battery life cycles.	• Specific regulations for second-life batteries are still developing.	• N/A
Identified lacks	• Lack of clear incentives for battery reuse and state regulatory differences. • Need for specific regulations and standards for second-life batteries. • Economic incentives for second-life applications are lacking.	• Need for business models for reuse and recycling. • Standards for second-life performance and safety may need further development.	• Lack of comprehensive legislation on battery lifecycle management.	• Lack of regulations of LIBs, in particular, with a focus on reuse and recycling. • Development of recycling infrastructure and incentives. • It is necessary to improve battery collection and recycling systems.	• Definition of responsibilities for second-life batteries. • Harmonization of standards for second-life batteries is ongoing. • Economic models for second-life battery applications need refinement.	• Need for clearer guidelines on repurposing and safety standards.	• Need for greater clarity and infrastructure.

Source: Rufino Júnior et al. [5] / MDPI / CC BY 4.0.

Aspect	Europe	United States	China
Extended producer responsibility (EPR)	Yes, comprehensive EPR mandates that producers are responsible for the entire lifecycle of batteries.	Fragmented approach with state-level regulations; EPR principles are applied in some states.	Yes, EPR principles are applied, but the system is less transparent.
Regulatory framework	A stringent and detailed legislative framework emphasizing sustainability and circular economy practices.	A more flexible and variable approach, relying on a combination of state and federal guidelines.	Centralized regulatory system with significant progress in establishing standards and policies for battery recycling and reuse.
Recycling targets	Specific targets are set, such as requiring 50% of the weight of batteries to be recycled.	Lack of strict, nationwide standards for recycling efficiency; some states prohibit landfill disposal of lithium-ion batteries.	Clear targets for battery recycling, focusing on efficiency and recovery of valuable materials.
Technical standards	Emphasis on safety, performance, and sustainability; compliance with international standards like UL 1974.	Compliance with technical standards for safety and performance; less emphasis on harmonized national standards.	Development of national standards for battery safety and performance; significant focus on recycling technologies.
Legislative gaps	Need for clearer definitions of responsibilities for battery ownership and recycling processes.	Fragmentation and lack of strict standards can lead to inconsistencies in battery management.	Less transparency in regulatory processes; need for improved infrastructure for battery collection and recycling.
Technological and infrastructure barriers	Advanced infrastructure for battery collection and recycling; focus on improving recycling technologies.	Developing infrastructure with challenges in the collection, sorting, and recycling of batteries.	Significant infrastructure for battery production and recycling; need for technological advancements in recycling processes.
Harmonization of standards	Efforts toward global harmonization to facilitate international trade and compliance.	Need for more consistent national standards to reduce market uncertainty and promote sustainability.	Progress in establishing national standards; potential for global harmonization to improve international trade.
Data sharing and transparency	Initiatives to encourage data sharing and development of digital Battery Passports for traceability and safety.	Challenges in standardized data sharing practices; need for improved transparency in battery health and history.	Efforts to improve data sharing and transparency in battery lifecycle management.

The lack of harmonization is particularly problematic in a supply chain as internationalized as that of batteries, where raw materials mined in Africa may be refined in Asia, assembled in Europe, and consumed in North America. Without interoperable systems, digital traceability breaks down at regional borders, defeating the purpose of lifecycle accountability. In some regions, such as Latin America and Africa, comparatively underdeveloped infrastructure and regulatory mechanisms are a barrier to full participation in global systems. Meanwhile, in the United States, the second Trump administration's protectionist stance poses challenges to international cooperation, with evolving policy positions contributing to an atmosphere of uncertainty for global markets.

If global stakeholders are serious about building a just and sustainable battery ecosystem, then establishing a shared framework aligned on core principles of traceability, environmental stewardship, and digital interoperability is essential. The path forward must prioritize international dialogue, mutual recognition of standards, and targeted support to bridge capacity gaps, especially in emerging economies. Until then, the fragmented nature of global Battery Passport efforts will continue to limit their collective impact, threatening to turn a powerful tool for sustainability into a patchwork of disconnected regional experiments.

4.8.4 Impact of this Policy Fragmentation

4.8.4.1 Economic Implications of Policy Difference

4.8.4.1.1 Increased Compliance Costs and Market Barriers Fragmented policies across regions result in varying compliance requirements, imposing significant financial and operational burdens on companies. For instance, battery manufacturers navigating both EU and U.S. markets must adhere to distinct lifecycle reporting obligations and data formats. These discrepancies increase administrative expenses and require firms to allocate additional resources for compliance.

Smaller companies, in particular, face disproportionately high barriers to market entry due to the lack of resources to navigate complex and inconsistent regulatory environments. Over time, these hurdles may reduce competitiveness and innovation, limiting the widespread adoption of Battery Passports [5].

4.8.4.1.2 Hindrance to International Trade Regional policy fragmentation often functions as a nontariff barrier, complicating the global trade of batteries and their components. Firms must tailor their production processes or redesign products to meet region-specific regulations, which delays market entry and increases production costs. For example, differences in data reporting standards between China, the EU, and the United States can create redundancies that slow down supply chain operations.

This lack of harmonization undermines the potential of Battery Passports as a tool for seamless international trade. Companies may struggle to scale operations globally, limiting their ability to meet rising demand for sustainable batteries [12].

4.8.4.1.3 Potential Benefits of Global Policy Coordination Global efforts to align Battery Passport regulations can address these challenges by establishing consistent standards. Harmonized policies enable companies to streamline operations, reduce compliance costs, and simplify international trade. By fostering collaboration, standardized frameworks can unlock efficiencies in global battery production and recycling, benefiting both the environment and the economy.

For example, a unified approach to reporting lifecycle emissions and material sourcing would not only facilitate compliance but also help countries meet shared environmental goals. The adoption of such standards could make Battery Passports more economically viable, encouraging broader participation from industry stakeholders (Figure 4.2) [14].

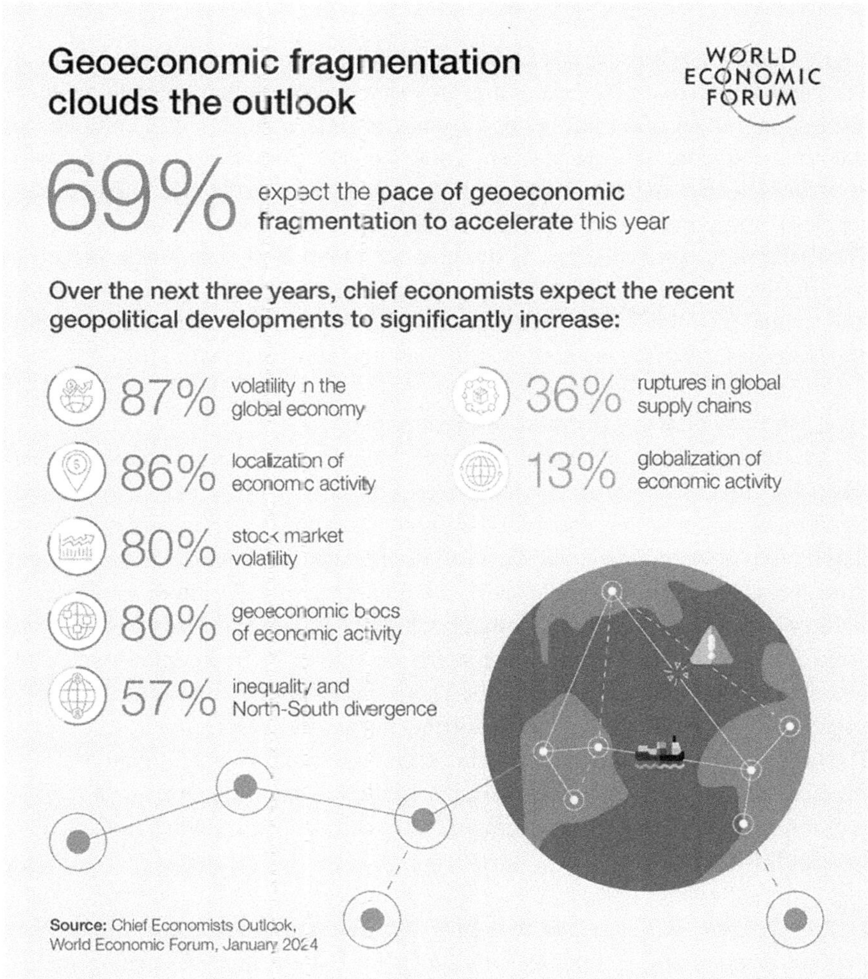

Figure 4.2 Geoeconomic fragmentation. *Source:* Karunska and Feingold [13] / World Economic Forum.

4.8.4.1.4 Opportunities for Lowering Barriers to Market Entry Harmonized global standards can significantly lower entry barriers for smaller firms and emerging markets. By simplifying regulatory landscapes, coordinated policies enable companies to expand into new markets without incurring prohibitive compliance costs. This fosters innovation, competition, and diversity in the global battery sector.

DPPs can significantly enhance transparency and sustainability within the EV battery industry by providing reliable information across the value chain. By standardizing data collection and sharing, DPPs enable companies to focus on improving product performance and environmental outcomes. This transparency fosters innovation and competition, driving the adoption of sustainable practices and creating a more competitive market for battery-related technologies. Ultimately, DPPs support the circular economy by facilitating the reuse, repair, and recycling of battery components, contributing to a more sustainable and efficient industry [15].

4.8.4.1.5 Balancing Economic and Environmental Goals Global policy harmonization offers the potential to balance economic growth with environmental sustainability. Coordinated frameworks promote consistency in data collection and compliance procedures, reducing redundancies and trade barriers. For example, aligning policies on lifecycle tracking or carbon footprint disclosures ensures that companies can meet global sustainability goals without sacrificing market competitiveness.

This balanced approach supports the broader adoption of sustainable practices across the battery value chain. By integrating environmental and economic priorities, policy coordination helps ensure that Battery Passports drive meaningful progress toward a circular economy while enabling industry growth [16].

4.8.4.2 Technological and Operational Consequences

4.8.4.2.1 Incompatibility of Digital Systems Policy fragmentation across global markets has direct technological implications, particularly in the development and integration of digital Battery Passports. Without a harmonized framework, regional systems often rely on incompatible formats, metadata schemas, or blockchain architectures. For example, while the EU is investing in fully interoperable data structures aligned with ISO/IEC standards, China's national platform is largely centralized and state-managed, and U.S. systems are shaped by fragmented state-level approaches and private-sector standards. This incompatibility undermines the very foundation of Battery Passports: seamless tracking across borders.

The lack of a common language for data exchange results in operational friction. Manufacturers and recyclers must maintain multiple systems, convert data manually, or rely on intermediaries for interoperability, each step introducing inefficiency, delays, and risk of error.

4.8.4.2.2 Delayed Innovation and Scalability Fragmented policies also hinder the scalability of emerging technologies designed for battery lifecycle management. Companies developing AI-based diagnostics, second-life battery monitoring tools, or blockchain-enabled traceability platforms face difficulty deploying them across markets due to incompatible policy and reporting requirements. This slows down innovation and increases the cost and risk of market entry for tech-driven startups and SMEs.

Additionally, second-life battery use cases such as repurposing EV batteries for stationary storage are slowed by unclear or inconsistent regulations, especially regarding safety certification and liability. In the absence of harmonized guidelines, technical innovation outpaces regulatory clarity, creating a bottleneck that restrains commercial adoption and slows circular economy progress.

4.8.4.3 Environmental and Circular Economy Trade-Offs

4.8.4.3.1 Missed Opportunities for Resource Optimization Battery Passports are designed to advance the circular economy by enabling the reuse, repurposing, and recycling of batteries. However, fragmented regulations compromise the efficiency of this loop. Diverging requirements for end-of-life classification, recycling thresholds, or reuse eligibility create conflicting incentives. For instance, a battery deemed fit for repurposing in China might be classified as hazardous waste in the United States, preventing its reuse in certain jurisdictions.

This regulatory misalignment reduces material recovery rates, increases the carbon intensity of recycling operations (due to redundant transport or reprocessing), and leads to avoidable resource waste. Rather than creating a closed-loop economy, fragmentation sustains a linear extraction–use–discard model, especially in emerging markets with limited capacity to adapt to multiple frameworks.

4.8.4.3.2 Environmental Leakage Through Regulatory Arbitrage Uneven regulatory enforcement creates opportunities for companies to engage in "regulatory arbitrage," shifting manufacturing or end-of-life operations to jurisdictions with weaker environmental or labor standards. This undercuts global sustainability efforts and allows batteries to escape accountability mechanisms, undermining the Battery Passport's role as a compliance tool.

Countries with underdeveloped e-waste enforcement, such as parts of Latin America or Sub-Saharan Africa, risk becoming dumping grounds for noncompliant or undocumented batteries. This contributes to environmental degradation, public health risks, and reputational damage to manufacturers operating across these fragmented ecosystems.

4.8.4.4 Political Uncertainty and Geopolitical Risk

4.8.4.4.1 Unpredictable Policy Shifts In regions where policy cycles are closely tied to political changes, such as the United States, the lack of legislative continuity introduces further instability. At the time of writing, the Trump administration has already taken steps to end EV subsidies and other climate-forward initiatives introduced just a few years ago under the Biden administration. Tensions stemming from the new administration's protectionist stance have introduced volatility into financial markets and heightened uncertainty around key trade relationships. This deters long-term investment and forces multinational firms to prepare for divergent regulatory futures.

The broader geopolitical rivalry between the United States and China adds another layer of complexity. Competing standards may emerge not just from differing environmental philosophies but from efforts to assert technological sovereignty and block market access. Rather than fostering interoperability, geopolitical tension may incentivize "splinter systems" where regional standards become tools of economic leverage rather than environmental governance.

4.8.4.4.2 Risk of Trade Fragmentation and Digital Protectionism As battery data becomes a strategic asset essential for resource planning, trade compliance, and environmental, social, and governance (ESG) reporting, the risk of digital protectionism grows. Countries may limit the cross-border transfer of battery lifecycle data, citing cybersecurity, national competitiveness, or privacy. This creates silos of information that limit the ability of supply chain actors to verify compliance, corroborate sustainability claims, or access secondary markets for recycled materials.

In this scenario, global battery value chains become increasingly segmented, slowing the transition to clean energy technologies that rely on shared infrastructure and knowledge exchange. Policy fragmentation thus fuels a larger risk: the splintering of environmental governance itself into incompatible blocs, with parallel standards that inhibit cooperation and slow progress.

4.8.4.5 Strategic Implications for the Global Battery Ecosystem

4.8.4.5.1 Market Consolidation and Inequity Larger corporations are better equipped to absorb the cost of operating across fragmented systems. This gives multinational firms a competitive edge, pushing out smaller companies and stifling regional innovation. The result is market consolidation, where a few dominant players shape standards and access, potentially limiting diversity in technologies and business models.

For developing countries, this dynamic may widen the sustainability divide. With limited capacity to implement complex tracking systems or meet evolving standards, these regions risk exclusion from future battery value chains either as suppliers or recyclers. If global Battery Passport systems are not designed with inclusivity in mind, the transition to a circular economy could reproduce the same extractive inequalities that it seeks to resolve.

4.8.4.5.2 Strategic Call for Action The danger is clear: policy fragmentation does not merely inconvenience industry actors. It threatens the integrity, sustainability, and inclusivity of the entire global battery ecosystem. Without unified regulatory architecture, Battery Passports may fail to deliver on their core promises of lifecycle accountability, resource efficiency, and environmental justice.

Therefore, the next phase of development must center on international alignment through bilateral agreements, multilateral standards, and global governance platforms. The cost of continued fragmentation is not only economic inefficiency but also the missed opportunity to build a truly sustainable and resilient battery future.

4.9 Solutions for the Fragmentation and Battery Passport Uptake

4.9.1 Harmonizing Global Standards for Battery Lifecycle Management

The transition to a circular economy for batteries depends heavily on the harmonization of global standards and regulatory frameworks. While different regional actors have made strides in developing policies for battery lifecycle management, significant disparities remain. The EU has emerged as a leader with its proposed Battery Passport framework, emphasizing traceability, lifecycle transparency, and sustainable material use. Meanwhile, North America and other regions face challenges due to fragmented regulations and the absence of centralized frameworks for lifecycle accountability [5].

A key driver for global harmonization is the interconnected nature of supply chains. Regions such as North America and Africa supply significant volumes of raw materials for batteries. Under the EU's developing regulations, these materials must meet stringent traceability and environmental standards. This interconnectedness underscores the need for a unified global approach to battery management that addresses shared challenges like recycling inefficiencies, transport logistics, and sustainable design [5].

This section explores how global standards must align to support a sustainable circular economy, emphasizing the importance of EU leadership and the ripple effects on other regions, particularly North America. It also examines the challenges and opportunities presented by harmonization efforts, including the implications for raw material suppliers and the evolving role of EPR.

4.9.1.1 EU Leadership in Driving Global Standards

The EU's legislative efforts, including its proposed Battery Passport initiative, serve as a benchmark for global battery lifecycle management. This system requires detailed documentation of a battery's composition, carbon footprint, and sourcing of raw materials throughout its lifecycle. The primary goal is to create a transparent supply chain that promotes sustainability and discourages practices that harm the environment or exploit resources unethically [5].

By implementing mandatory regulations under its EPR framework, the EU ensures that producers remain accountable for batteries even after they leave the market. This includes obligations for proper recycling and reuse, which are key to fostering a circular economy. The EU's recycling targets, requiring 50% of a battery's weight to be recycled, are among the most ambitious globally and are supported by mechanisms such as collection fees and material recovery rates [5].

As these regulations gain traction, their influence extends beyond EU borders. Raw materials from North America destined for the EU market will need to comply with these standards. This dynamic is already pushing North American suppliers and manufacturers to adopt similar practices, ensuring traceability and lifecycle transparency to maintain access to the EU market.

4.9.1.2 North America's Challenges and Alignment Opportunities

Unlike the EU, North America lacks a unified regulatory framework for battery lifecycle management. The region's approach is largely decentralized, with varying policies at federal and state levels in the United States and a reliance on voluntary, market-driven initiatives. While this flexibility encourages innovation, it creates inconsistencies and gaps in lifecycle accountability [5].

One key challenge lies in the absence of comprehensive EPR policies in North America. Without clear mandates, responsibility for battery end-of-life management often falls on consumers or smaller operators, resulting in inefficiencies and environmental risks. For instance, the lack of standardized recycling infrastructure and inconsistent battery labeling make it difficult to achieve high recovery rates, particularly for LIBs. These challenges are compounded by high transportation costs, limited safety protocols for battery storage, and insufficient coordination across borders [5].

However, the EU's leadership offers an opportunity for North America to align its practices with emerging global standards. By adopting elements of the EU framework, such as Battery Passports or stricter recycling targets, North America could improve transparency and lifecycle accountability. Collaboration through trade agreements like the USMCA could further facilitate cross-border integration and drive the development of harmonized policies [5].

4.9.1.3 The Role of Raw Materials and EPR in Harmonization

Raw materials play a pivotal role in the battery supply chain, and the lack of global standards for their sourcing and traceability presents a major barrier to harmonization. North America's position as a key supplier of critical minerals like lithium, cobalt, and nickel makes it uniquely positioned to influence global practices. However, these materials must meet the traceability and sustainability requirements set by markets like the EU, creating pressure on North American suppliers to align with EU regulations [5].

The adoption of EPR frameworks could bridge this gap by ensuring that producers remain responsible for the lifecycle of their products, from sourcing to disposal. In the EU, EPR policies require manufacturers to bear the financial and operational burden of recycling, while in North America, such policies remain limited to specific states or voluntary initiatives. Expanding EPR mandates across North America could incentivize sustainable practices and reduce regulatory uncertainty for global manufacturers [5].

4.9.1.4 Challenges to Harmonizing Standards

Harmonizing global standards is not without its challenges. Disparities in regulatory frameworks, such as the EU's centralized approach versus North America's decentralized structure, create obstacles to alignment. Moreover, the classification of batteries as hazardous waste under the Basel Convention adds complexity to cross-border transportation, particularly for raw materials and end-of-life batteries [5].

Technological barriers, such as the lack of standardization in battery design and labeling, further complicate harmonization efforts. For example, LIBs often vary in composition and configuration, making automated disassembly and recycling inefficient. Addressing these challenges will require international collaboration to develop shared technical standards and invest in infrastructure for battery tracking and recycling [5].

4.9.1.5 Opportunities for Collaboration and Innovation

Despite these challenges, the push for harmonized global standards presents significant opportunities for innovation and collaboration. Digital tools such as blockchain and IoT-based tracking systems could facilitate traceability and compliance with evolving regulations. These technologies can provide real-time data on battery health, composition, and lifecycle status, improving transparency across the supply chain [5].

Additionally, the development of shared recycling targets and recovery rates could reduce regional disparities and foster a more integrated circular economy. For example, aligning North America's recycling practices with the EU's 50% recovery target could streamline compliance and improve resource efficiency. Collaborative efforts between governments, industry stakeholders, and international organizations could further accelerate progress toward harmonization [5].

Harmonizing global standards for battery lifecycle management is essential to achieving a sustainable circular economy. The EU's leadership, through its ambitious Battery Passport framework and EPR mandates, provides a roadmap for other regions, including North America, to align their policies with emerging global benchmarks.

While challenges such as regulatory disparities, technological barriers, and high costs remain, the interconnected nature of global supply chains creates a strong incentive for collaboration. For regions like North America, aligning with EU standards offers a pathway to maintain competitiveness and ensure access to international markets. By adopting unified policies and leveraging innovative technologies, the global battery industry can address shared challenges and advance sustainability across the supply chain [5].

Harmonizing battery passport standards globally could alleviate these challenges by providing a consistent framework for compliance. A unified approach would reduce administrative burdens for companies, enhance interoperability, and promote fair competition across markets [17]. *By fostering collaboration among governments, industry leaders, and standardization organizations, these efforts support the creation of a global ecosystem for sustainable battery management. Such cooperation is critical for ensuring that regional policy variations do not hinder progress toward achieving a circular economy and advancing sustainable trade* [5].

Government incentives play a crucial role in promoting the adoption of the Battery Passport system by reducing financial barriers and encouraging stakeholders to comply with new regulatory standards. Incentive programs such as tax breaks, subsidies, and grants can help accelerate the adoption of Battery Passports, promoting transparency, sustainability, and circularity within the battery supply chain.

4.9.2 The Role of Regional Trade Agreements

Regional trade agreements, such as the USMCA and the EU's Free Trade Agreements, play a pivotal role in advancing Battery Passport implementation. By embedding provisions for environmental sustainability, these agreements harmonize regulatory standards, streamline cross-border trade, and support the global shift toward a circular economy.

4.9.2.1 Harmonizing Regulatory Frameworks

One of the critical contributions of trade agreements is their ability to harmonize regulatory frameworks across member countries. Agreements like the USMCA and EU trade pacts include provisions for environmental sustainability, which align member states' Battery Passport policies. This harmonization reduces discrepancies between regulatory requirements, easing compliance for manufacturers operating across multiple regions. For instance, shared standards for data reporting and lifecycle tracking ensure consistency, saving companies time and resources.

4.9.2.2 Facilitating the Movement of Battery Components and Materials

One of the critical contributions of trade policies, such as those influenced by the Inflation Reduction Act (IRA), is their potential to harmonize regulatory frameworks across regions. The IRA emphasizes the need for consistent standards and regulatory approaches, particularly in the battery industry. By promoting a "Battery Passport" system, these policies aim to enhance traceability and ensure compliance with environmental and safety standards across borders. This harmonization can help reduce discrepancies between regulatory requirements, easing compliance for manufacturers operating across multiple regions and fostering a more integrated global battery value chain [18].

For example, under the USMCA, streamlined provisions for environmental cooperation enable easier tracking of battery materials across North America, supporting regional manufacturing and recycling initiatives.

4.9.2.3 Supporting Environmental Goals Through Trade Provisions

Many trade agreements explicitly link environmental goals with trade provisions, encouraging sustainable practices in member countries. Clauses related to carbon reduction, resource recovery, and recycling align with Battery Passport initiatives, incentivizing companies to adopt policies that prioritize circular economy principles. By fostering cross-border collaboration on sustainability, trade agreements amplify the environmental benefits of Battery Passports, driving progress toward global climate goals.

4.9.2.4 Mitigating Trade Barriers Through Policy Alignment

Inconsistent regulations between countries can create **nontariff barriers**, complicating compliance for companies operating internationally. Trade agreements address these challenges by aligning regional policies, mitigating regulatory disparities, and ensuring smoother integration of Battery Passports into global trade. For instance, EU trade pacts often require harmonized reporting on lifecycle emissions and material sourcing, which facilitates trade while maintaining high sustainability standards.

4.9.2.5 Integrating Environmental Standards with Trade Agreements

Trade agreements provide a platform for embedding environmental standards into broader trade policies. For instance, agreements that incorporate Battery Passport requirements into their frameworks establish clear expectations for compliance, encouraging companies to invest in environmentally friendly technologies and processes. This integration not only supports sustainability but also drives innovation, making the adoption of Battery Passports a strategic priority for companies seeking long-term competitiveness.

Regional trade agreements play a crucial role in bridging the gap between diverse policy frameworks, enabling the global adoption of Battery Passport systems. By harmonizing standards, facilitating cross-border trade, and integrating environmental goals, these agreements reduce barriers and foster collaboration. Moving forward, expanding the scope of trade agreements to include stricter sustainability criteria and digital technologies could further enhance the effectiveness of Battery Passports in achieving both economic and environmental objectives.

4.10 Further Challenges

Battery Passport regulations present various legal and ethical challenges, including concerns over data privacy, intellectual property rights, and competition law disputes. As companies adapt to these evolving regulations, they must navigate potential legal conflicts while adhering to ethical standards for sustainability and consumer protection.

4.10.1 Legal Challenges

The legal complexities of Battery Passports primarily stem from handling sensitive information, adhering to data protection laws, and safeguarding intellectual property. These challenges highlight the need for clear regulatory guidelines that can reconcile transparency with confidentiality.

4.10.1.1 Data Privacy Concerns

Battery Passports require extensive data collection and sharing across the supply chain, raising significant concerns over data privacy. In regions like Europe, compliance with stringent regulations such as the GDPR compounds the challenge, as companies must implement robust systems to prevent data misuse or breaches.

The integration of Battery Passports into existing data protection frameworks can lead to regulatory conflicts. For instance, while transparency demands open data sharing, privacy protections impose restrictions on how that data can be stored and used. Companies often find themselves navigating these opposing priorities, requiring innovative solutions that safeguard data while maintaining compliance.

4.10.1.2 Intellectual Property Issues

One of the most pressing legal challenges is balancing transparency with the protection of proprietary information. While Battery Passports aim to promote accountability by disclosing details about material provenance, lifecycle data, and carbon footprints, companies may be reluctant to share information that could reveal trade secrets or competitive advantages [19].

The issue becomes more pronounced with second-life batteries. Data on battery performance, condition, and history often involves proprietary technologies, which raises questions about ownership rights and the use of patented processes. For companies invested in innovation, the risk of exposing intellectual property through mandated data sharing can create resistance to Battery Passport adoption.

4.10.1.3 Competition Law Considerations

Battery Passport regulations also intersect with competition law, particularly when data-sharing requirements disproportionately impact smaller companies. While larger corporations can absorb compliance costs, SMEs may struggle to meet extensive reporting demands, leading to competitive disadvantages. Ensuring fairness in regulatory frameworks is critical to avoid stifling innovation and creating monopolistic dynamics within the battery sector.

Moreover, the enforced sharing of sensitive information among stakeholders can inadvertently facilitate anti-competitive practices. Regulatory bodies must carefully structure requirements to prevent misuse of data while fostering collaboration and innovation across the supply chain.

4.10.1.4 Legal Disputes and Resolutions

Disputes related to Battery Passports frequently arise over data privacy and intellectual property concerns. These conflicts are often resolved through a combination of policy adjustments, industry negotiations, and legal rulings aimed at balancing regulatory goals with business interests.

For example, when disputes over data privacy emerge, companies often negotiate data-sharing agreements to define how sensitive information will be stored, shared, and protected. Regulatory bodies may also issue clarifying guidelines to delineate the boundaries between data transparency and confidentiality.

Battery Passport regulations present a complex interplay of legal and ethical challenges. As the system evolves, companies must address issues related to data privacy, intellectual property, and fair competition. Resolving these challenges will require innovative solutions, collaborative negotiations, and clear regulatory frameworks that balance the need for transparency with the protection of proprietary and sensitive information. By navigating these hurdles, the Battery Passport system can achieve its potential as a transformative tool for sustainability and accountability.

4.10.2 Ethical Considerations in Policymaking

The implementation of Battery Passports involves critical ethical considerations, focusing on promoting transparency, sustainability, and consumer rights while ensuring that businesses are not unduly burdened by compliance requirements. Ethical policymaking plays a pivotal role in balancing economic growth with environmental protection, fostering accountability throughout the battery supply chain.

4.10.2.1 Transparency and Sustainability

At the core of Battery Passport initiatives lies the commitment to providing ethical transparency. Battery Passports enable consumers and stakeholders to access detailed information about the environmental and social impacts of batteries across their lifecycle. By highlighting responsible sourcing, ethical mining practices, and adherence to circular economy principles, these systems promote sustainability and fairness within the industry.

The ability to track compliance with ethical standards through Battery Passports reinforces accountability. For example, the extraction of raw materials such as cobalt and lithium, often associated with controversial mining practices, can be monitored to ensure responsible sourcing and compliance with environmental safeguards during manufacturing. This transparency not only helps mitigate unethical practices but also fosters trust among consumers and stakeholders.

4.10.2.2 Balancing Economic and Environmental Goals

Ethical considerations in Battery Passport policymaking must address the tension between environmental goals and economic competitiveness. Regulations like the EU Battery Regulation emphasize sustainability while acknowledging the necessity for manufacturers to remain competitive in global markets. Striking this balance ensures that environmental protections do not stifle innovation or impose unsustainable costs on businesses.

One way to achieve this balance is through initiatives that combine environmental objectives with economic strategies. For instance, climate-smart public procurement practices, such as those implemented by the city of Malmö, demonstrate how ethical policies can achieve sustainability goals without compromising economic viability. Malmö's roadmap for integrating ethical and climate-conscious procurement practices highlights how Battery Passports can facilitate environmentally responsible decision-making while promoting efficiency and cost-effectiveness.

4.10.2.3 The Role of Consumer Rights in Ethical Policymaking

Battery Passport systems also address ethical concerns around consumer rights, ensuring that individuals have access to transparent and accurate information about the products they use. By providing data on carbon footprints, material sourcing, and recyclability, Battery Passports empower consumers to make informed decisions, aligning with broader principles of environmental justice and accountability. Policies that prioritize consumer rights contribute to fostering trust between manufacturers and their end users, reinforcing the ethical dimensions of battery production and use.

Ethical considerations are at the heart of Battery Passport implementation, driving transparency, accountability, and sustainability within the industry. By addressing challenges such as responsible sourcing, economic feasibility, and consumer rights, policymakers can create frameworks that balance environmental objectives with business innovation and competitiveness. Initiatives like climate-smart procurement and transparent lifecycle reporting illustrate how ethical policymaking can support long-term sustainability while promoting fairness and trust across the battery value chain.

4.10.3 Compliance Challenges

While government incentives facilitate the adoption of the Battery Passport system, compliance remains a significant challenge for stakeholders. Companies must navigate stringent and evolving regulatory requirements, ensure data accuracy and security, and adapt to cross-border trade complexities. These issues are compounded by the need to meet recycling targets, adopt new technologies, and harmonize standards across regions.

4.10.3.1 The Role of Trade Agreements in Streamlining Compliance

Trade agreements such as the USMCA and the CPTPP play a critical role in aligning regulations across borders. By harmonizing standards, these agreements reduce compliance barriers for multinational companies, fostering smoother adoption of Battery Passports and ensuring consistency in sustainability goals.

4.10.3.2 Challenges in the Asia-Pacific Region

The Asia-Pacific region faces unique hurdles due to its highly diverse regulatory environment. Countries like Japan and South Korea have advanced standards for battery recycling, while ASEAN nations often lack the infrastructure and policy coherence needed for effective implementation. Harmonizing these regulations is essential for ensuring Battery Passports can function effectively throughout the region.

4.10.3.3 Data Privacy

Manufacturers face risks associated with sharing proprietary data, including potential breaches or misuse by third parties. In regions like Europe, compliance with laws such as the GDPR adds complexity to managing sensitive information securely. The cost and administrative burden of meeting diverse data protection regulations globally are significant.

Data privacy challenges are exacerbated by varying legal frameworks across jurisdictions. For instance, China's Data Security Law and the U.S.'s California Consumer Privacy Act (CCPA) impose different obligations, increasing costs for global compliance. Developing common data-sharing protocols can reduce these burdens while maintaining confidentiality. An example of a similar struggle was the rollout of Apple Intelligence that took close to a year to arrive in Europe.

4.10.3.4 Reporting Requirements

Granular reporting requirements place a disproportionate burden on smaller companies, which often lack the resources to maintain sophisticated data systems. This can significantly increase operational costs and administrative workloads, potentially hindering compliance and competitiveness [20].

Streamlining reporting templates and setting international data-sharing standards can alleviate these challenges. Simplified compliance processes, supported by unified protocols, can help SMEs meet regulatory requirements more efficiently.

4.10.3.5 Challenges for SMEs

Adopting digital solutions for Battery Passports requires significant investment in IT infrastructure, data management, and interoperable systems. Real-time data tracking and compliance monitoring pose particular challenges for SMEs, which may lack the financial and technical resources needed for implementation.

Governments can mitigate these barriers by offering targeted financial support for SMEs' digital investments, such as consultancy vouchers and grants, and by using subsidies and tax incentives to lower adoption and training costs. Public–private partnerships and ecosystem mechanisms such as digital innovation hubs, testbeds and experimentation labs, and co-creation platforms can further help SMEs access external expertise and affordable digital solutions without building costly capabilities in-house [21].

4.11 Future Policy Trends and Directions

4.11.1 Implementation Challenges Across the Value Chain

While the development of Battery Passport systems may be finalized on schedule, their practical implementation across the global value chain could face significant delays due to the inherent complexity of the industry. One major challenge is the diversity of stakeholders involved in the battery lifecycle, including raw material suppliers, manufacturers, EV producers, recyclers, and logistics providers. These actors operate at varying levels of technological and organizational readiness, with smaller enterprises or stakeholders in developing regions often lacking the resources to integrate advanced digital tools. Aligning such a diverse group under a unified system is likely to be a time-consuming process.

The global nature of supply chains further compounds these challenges. Batteries and their raw materials frequently cross multiple jurisdictions with differing regulatory requirements, economic priorities, and technological capabilities. Harmonizing these fragmented processes to comply with a uniform Battery Passport framework will require substantial coordination, particularly when existing systems are incompatible across regions.

Mining operations present a unique challenge in meeting Battery Passport requirements. Many mines, especially in developing nations, operate with outdated technologies and limited oversight. Implementing traceability and sustainability measures will require substantial investment in monitoring equipment, data collection systems, and workforce training. These operations may also face logistical challenges, such as the remote location of mines, poor infrastructure, and a lack of access to advanced digital tools like blockchain. Furthermore, sociopolitical issues, such as governance challenges, corruption, or the prevalence of informal and unregulated mining practices, can make compliance nearly impossible in certain regions. As a result, many mining operations will be unable to meet the stringent traceability and reporting requirements of Battery Passports for years, creating a bottleneck at the very start of the value chain.

Technology, while a key enabler, is another source of potential delays. Many of the proposed Battery Passport policies rely on transformative tools such as blockchain, IoT, and digital platforms. However, the uneven global deployment of these technologies creates bottlenecks, as some regions or actors may lack the infrastructure or expertise to adopt them seamlessly. Additionally, data collection and verification central to the success of Battery Passports require robust systems to manage extensive information flows. Ensuring the security, accuracy, and interoperability of this data will be an enormous logistical and technological undertaking.

Another critical factor is the lack of standardization across regions. While the EU has made strides in defining its Battery Passport framework, other regions like North America and Asia operate under disparate systems or voluntary guidelines. Achieving global standardization will require extensive negotiation and alignment, further delaying implementation timelines. Supporting infrastructure, such as recycling facilities and battery collection systems, also remains underdeveloped in many regions, limiting the ability to operationalize policies effectively.

Resistance to change within the industry could exacerbate these delays. Implementing Battery Passports will require significant investments from stakeholders, which may face pushback in cost-sensitive sectors or in regions with limited enforcement mechanisms. Additionally, training the workforce across the value chain to effectively navigate new technologies and regulatory requirements will require considerable time and effort.

Economic and political dynamics could further complicate the rollout. Geopolitical tensions, economic crises, or competing policy priorities in key regions may deprioritize sustainability initiatives, delaying funding and enforcement. Combined with challenges in compliance and governance, these factors could create a gap between the establishment of Battery Passport systems and their full implementation.

4.11.1.1 The Resulting Gap Between Policy and Practice

These complexities suggest that full-scale implementation may lag significantly behind policy and system development. This delay could be particularly acute at the mining stage, where the adoption of traceability measures is hindered by technological, economic, and governance barriers. Such delays could undermine the immediate impact of these policies on global sustainability efforts, requiring sustained investment and international cooperation to bridge the gap between policy and practice. Addressing these challenges will require not only technological advancements but also robust governance frameworks and strategic incentives to ensure that all stakeholders, including mining operations, are aligned and prepared to adopt Battery Passports effectively (Figure 4.3).

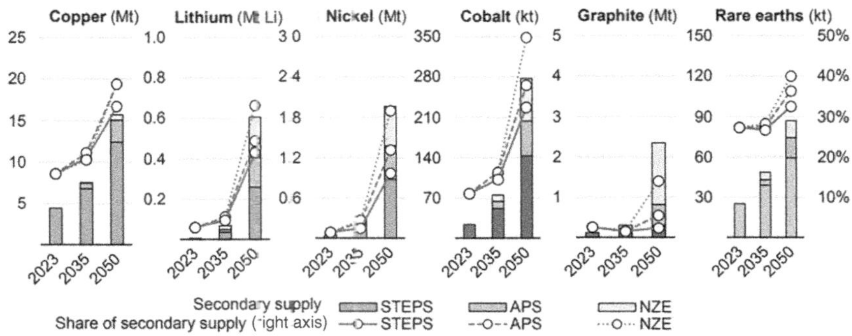

Figure 4.3 Secondary supply volumes and share of total demand for key energy transition minerals in the APS and NZE scenario, 2023–2050. Notes: Mt, million tons; kt, kilotons. Includes recycled volumes from end-of-life equipment and manufacturing scrap. For copper, direct use scrap is excluded. For nickel, scrap from the steel sector is excluded. For rare earth elements, the figures are for magnet rare earth elements only (neodymium, praseodymium, dysprosium, and terbium). *Source:* Recycling of Critical Minerals IEA / IEA / CC BY 4.0.

4.11.2 Adapting to Future Policy Developments

As the regulatory landscape surrounding Battery Passports continues to evolve, stakeholders must employ proactive strategies to navigate new challenges and capitalize on emerging opportunities. Adapting effectively requires a combination of vigilance, innovation, and collaboration. Below are key approaches to ensure readiness for future policy developments:

4.11.2.1 Regular Monitoring of Policy Changes

Staying informed about regulatory updates is crucial to maintain compliance and a competitive advantage. Manufacturers, recyclers, and other stakeholders should consistently monitor both domestic and international developments in Battery Passport policies. Subscribing to industry newsletters, participating in trade groups, and attending conferences are essential strategies for staying abreast of regulatory changes. These channels not only provide updates on best practices but also offer insights into future trends and policy directions.

4.11.2.2 Building Flexibility into Business Models

To remain adaptable, companies should embed flexibility into their operational frameworks. This includes designing modular systems that can easily integrate new reporting and tracking requirements, such as those associated with updated Battery Passport standards. Collaborating with technology providers can help streamline compliance processes, enabling businesses to respond quickly to new regulations. By adopting a forward-thinking approach, companies can mitigate disruption and maintain efficiency amid regulatory shifts.

4.11.2.3 Engagement with Policymakers

Active participation in the policymaking process is critical for ensuring that industry perspectives are reflected in regulations. Stakeholders should engage with regulatory bodies through industry associations and advocacy groups, providing feedback during consultation phases or participating in pilot projects. These efforts can influence the development of practical, industry-aligned policies while building stronger relationships with regulatory authorities. Engagement also positions stakeholders as proactive participants in shaping the future of the battery industry.

4.11.2.4 Adopting Emerging Technologies

The adoption of emerging technologies is becoming increasingly vital for compliance with Battery Passport regulations. Blockchain can enhance supply chain transparency and ensure secure, tamper-proof data sharing, while AI can improve the accuracy of reporting and automate compliance processes. Investing in these technologies reduces administrative burdens and positions companies to meet regulatory standards in real time. As regulations evolve, early adoption of these tools will offer a competitive edge and facilitate seamless adaptation to new requirements.

Adapting to the dynamic regulatory landscape of Battery Passports requires vigilance, flexibility, and innovation. Regular monitoring of policy updates, building adaptable business models, engaging with policymakers, and leveraging emerging technologies are essential strategies for staying ahead. By proactively addressing these areas, stakeholders can navigate future policy developments with confidence, ensuring compliance while driving sustainability and innovation within the battery industry.

4.11.2.5 Acknowledging Implementation Challenges

While the framework for Battery Passports may be developed on time, the complexity of implementing these policies across a global value chain could lead to significant delays. Stakeholders must recognize that challenges such as fragmented supply chains, technological disparities, and uneven regulatory enforcement will affect the speed at which these policies are fully operationalized.

Outdated technologies, insufficient monitoring systems, and informal or unregulated mining practices further compound this challenge. Even with the support of advanced tools like blockchain or IoT, the adoption of traceability mechanisms across mining operations is expected to lag behind the broader supply chain.

These challenges highlight the need for realistic timelines and transitional policies that support capacity building, particularly in regions where compliance with Battery Passport standards is most difficult. Investments in training, infrastructure development, and stakeholder collaboration will be essential to bridging these gaps. By acknowledging these potential delays and actively planning for them, stakeholders can better align their strategies with the realities of implementation, ensuring that policies achieve their intended impact over time.

This emphasis on implementation readiness underscores the importance of flexibility and long-term commitment, positioning stakeholders to adapt effectively as Battery Passport systems move from policy design to real-world application.

4.12 Conclusions

The Battery Passport has the potential to advance sustainability across the battery value chain – if stakeholders can master the intricate interplay between global regulatory frameworks, government incentives, stakeholder collaboration, and technological advancements.

4.12.1 Strengthening the Global Framework with UNECE Regulations

The incorporation of UNECE regulations highlights the critical importance of embedding robust safety and sustainability standards into Battery Passports. These regulations address potential hazards such as thermal runaway, crash safety, and operational reliability, ensuring that batteries meet stringent international benchmarks. Their integration into the Battery Passport system not only enhances consumer trust and global alignment but also reinforces the passport's role as a tool for proactive risk management. By incorporating UNECE's sustainability-focused provisions, the Battery Passport system can better support the global transition to safe and environmentally responsible EV adoption.

4.12.2 The Role of Government Incentives in Driving Adoption

The insights into tax breaks, subsidies, and grants illuminate how government incentives serve as powerful catalysts for accelerating the implementation of Battery Passports. By addressing financial barriers and supporting technological upgrades, these programs enable stakeholders to adopt Battery Passport systems more seamlessly. Evidence from regions like the EU, China, and North America illustrates how targeted subsidies and grants encourage innovation, particularly in recycling technologies and second-life battery applications. This demonstrates that a carefully crafted incentive structure is pivotal in balancing compliance with financial feasibility, particularly for SMEs and emerging markets.

4.12.3 Addressing Compliance Challenges with Collaboration and Innovation

While government incentives create pathways for adoption, compliance challenges remain a significant hurdle. The regulatory complexity of meeting diverse standards across regions, coupled with data privacy concerns and resource limitations, highlights the need for harmonization and technological solutions. The chapter's insights into cross-border collaboration, facilitated by trade agreements and international bodies like the UNECE and ISO, underscore the importance of developing unified frameworks that reduce compliance burdens while maintaining transparency and accountability. Leveraging emerging technologies such as blockchain, AI, and IoT will further streamline compliance and enable real-time monitoring of battery lifecycle data, ensuring that stakeholders can adapt to regulatory demands efficiently.

4.12.4 Ethical and Legal Considerations: Balancing Transparency and Privacy

Stakeholders must maintain a delicate balance between promoting transparency and protecting proprietary information. Addressing data privacy concerns and intellectual property rights will require innovative governance models that ensure data security while fostering trust among stakeholders. Ethical considerations, such as ensuring responsible sourcing and prioritizing consumer rights, remain central to the design of Battery Passports. These dimensions highlight the system's dual role as both a regulatory tool and a mechanism for advancing social and environmental justice.

4.12.5 Anticipating Future Trends and Implementation Challenges

Emerging trends will shape the evolution of Battery Passport policies, including the integration of transformative technologies and the increasing emphasis on circular economy principles. While these advancements hold immense potential, they also present significant challenges, particularly in regions with fragmented supply chains, outdated infrastructure, and governance gaps. The unique challenges faced by mining operations, such as the lack of traceability mechanisms and technological disparities, underscore the need for transitional policies and capacity-building initiatives. By addressing these challenges proactively, stakeholders can bridge the gap between policy design and practical implementation, ensuring the long-term success of Battery Passport systems.

4.12.6 A Unified Vision for Sustainability

We reaffirm the Battery Passport's potential to transform the battery industry into a global model for sustainability, innovation, and ethical accountability. By integrating comprehensive regulatory frameworks with advanced technologies, government incentives, and international collaboration, the Battery Passport can catalyze systemic change across the value chain. However, its success depends on addressing implementation challenges, fostering stakeholder engagement, and harmonizing regional disparities.

The Battery Passport has the potential to be a foundational tool for worldwide sustainability efforts. By promoting transparency, reducing environmental impact, and supporting circular economy practices, it aligns the interests of manufacturers, recyclers, policymakers, and consumers with the broader goal of achieving global sustainability. The integration of UNECE regulations, government incentives, and transformative technologies further strengthens this vision, ensuring that Battery Passports remain at the forefront of efforts to create a cleaner, greener, and more equitable world. Through collective action and sustained commitment, stakeholders can ensure that the promise of Battery Passports is fully realized, delivering lasting benefits for the planet and future generations.

References

1 Megyeri, E. and Tabajdi, G. (2024). Recollecting, reusing, and recycling: the objectives and impacts of EU legislation on the battery value chain. *Business Logistics in Modern Europe* 91. https://www.researchgate.net/publication/384600285.

2 Berger, K., Schöggl, J.P., and Baumgartner, R.J. (2022). Digital Battery Passports to enable circular and sustainable value chains: conceptualization and use cases. *Journal of Cleaner Production* 353: 131492. https://www.sciencedirect.com/science/article/pii/S0959652622011131.

3 Lin, Y.C. (2023). Impact of the EU battery regulation on circular economy of EV batteries: barriers and opportunities for battery repurposing. https://www.diva-portal.org/smash/record.jsf?pid=diva2:1846184.

4 Liu, F., Tang, J., Rustam, A., and Liu, Z. (2023). Evaluation of central and local power battery recycling policies in China. *Journal of Cleaner Production*. https://www.sciencedirect.com/science/article/pii/S0959652623032316.

5 Rufino Júnior, C.A., Riva Sanseverino, E., Gallo, P., and Koch, D. (2024). Towards battery digital passport: reviewing regulations and standards for second-life batteries. *Batteries* 10 (4): 115. https://opus4.kobv.de/opus4-haw/files/4645/a.pdf.

6 Nurdiawati, A. and Agrawal, T.K. (2022). Creating a circular EV battery value chain: end-of-life strategies and future perspective. *Resources, Conservation and Recycling* 185: 106484. https://www.sciencedirect.com/science/article/pii/S0921344922003275.

7 Federal Consortium for Advanced Batteries (2021). *National Blueprint for Lithium Batteries 2021–2030*. U.S. Department of Energy. https://www.energy.gov/sites/default/files/2021-06/FCAB%20National%20Blueprint%20Lithium%20Batteries%200621_0.pdf.

8 Meegoda, J.N., Malladi, S., and Zayas, I.C. (2022). End-of-life management of electric vehicle lithium-ion batteries in the United States. *Clean Technologies* 4 (4): 1162–1174. https://www.mdpi.com/2571-8797/4/4/71.

9 Mususa, P. and Shengo, M. (2024). A battery industry in the Central African Copperbelt? Regional and geopolitical dimensions. https://www.researchgate.net/publication/384887664.

10 Lotter, A. (2024). Energising solutions for lithium-ion battery waste management in South Africa and India. https://www.sustainable-recycling.org/wp-content/uploads/2025/08/2024_Energising-Solutions-for-Lithium-Ion-Battery-Waste-Management-in-South-Africa-and-India_Lotter.pdf.

11 Barrie, J., Latif, L.A., and Albaladejo, M. (2022). Trade for an inclusive circular economy. Royal Institute of International Affairs. https://www.chathamhouse.org/sites/default/files/2022-06/2022-06-15-inclusive-circular-trade-barrie-et-al.pdf.

12 Milios, L. and Dávila, M.L. (2023). Digital product passports: implications of policy coherence and implementation towards a circular economy. https://www.researchgate.net/publication/376208532.

13 Karunska, K. and Feingold, S. (2024). What will happen to the global economy in 2024? Leading chief economists give their views [Internet]. World Economic Forum. https://www.weforum.org/stories/2024/01/economic-outlook-2024-recession-inflation/.

14 Durand, A., Goetz, T., Hettesheimer, T., and Ziegler, F. (2022). Enhancing evaluations of future energy-related product policies with the digital product passport. In: *Proceedings of Energy Evaluation Europe 2022 Conference*, 28–30. Paris-Saclay University. https://publica-rest.fraunhofer.de/server/api/core/bitstreams/144964c7-b5df-49c4-96fc-ed558e3076f1/content.

15 Kim, J., Moon, J., Do, Y., and Kim, H. (2023). Research on a SaaS (software as a service)-based digital product passport system model for the EV battery industry value chain. *WSEAS Transactions on Business and Economics* 20: 2466–2475. https://wseas.com/journals/bae/2023/e305107-037(2023).pdf.

16 Berger, K., Baumgartner, R.J., and Weinzerl, M. (2023). Data requirements and availabilities for a digital Battery Passport – a value chain actor perspective. https://www.sciencedirect.com/science/article/pii/S2666791623000052.

17 Moïsé, E. and Rubínová, S. (2023). Trade policies to promote the circular economy: a case study of lithium-ion batteries. OECD. https://www.oecd-ilibrary.org/trade/trade-policies-to-promote-the-circular-economy_d75a7f46-en.

18 Pedersen, E.L. (2023). Navigating the inflation reduction act: impacts on the battery industry, transatlantic trade, and green transitions. https://nmbu.brage.unit.no/nmbu-xmlui/handle/11250/3076810.

19 King, M.R.N., Timms, P.D., and Mountney, S. (2023). A proposed universal definition of a digital product passport ecosystem. *Journal of Cleaner Production*. https://www.sciencedirect.com/science/article/pii/S0959652622051125.

20 Orgalim aisbl (2025). Time to act: reducing the EU's regulatory burden on Europe's technology industries: 10-point action plan & practical recommendations for better EU regulation [Internet]. Brussels (Belgium): Orgalim aisbl. https://orgalim.eu/wp-content/uploads/EU_RegulatoryBurden-1.pdf.

21 Organisation for Economic Co-operation and Development (OECD) (2021). The digital transformation of SMEs: OECD policy highlights. Paris: OECD Publishing. https://www.oecd.org/content/dam/oecd/en/publications/support-materials/2021/02/the-digital-transformation-of-smes_ec3163f5/Digital-Transformation-of-SMEs-policy-highlights-en.pdf.

5

Addressing Greenhouse Gas Emissions and Closed-Loop Recycling in the Battery Industry: The Role of Battery Passports

The global transition to EVs and renewable energy storage systems has significantly increased the demand for lithium-ion batteries. While these batteries are pivotal for decarbonization, their lifecycle from raw material extraction to disposal poses severe environmental challenges. Key issues include high GHG emissions, resource depletion, and inadequate recycling practices, which threaten the sustainability of the clean energy transition. This chapter explores how Battery Passports, digital tools that track battery lifecycles, address these challenges by promoting transparency, circular economy principles, and accountability across the battery supply chain.

5.1 The Environmental Challenges of Battery Production and Disposal

5.1.1 Greenhouse Gas Emissions

Battery production is inherently carbon-intensive, from the extraction and processing of raw materials such as lithium, cobalt, and nickel, and the synthesis of cathode and anode active materials, to the production of battery cells; all contribute significantly to GHG emissions. A recent study from Roland Berger [1] suggests that the carbon footprint of a battery cell varies between 70 and 90 kg CO_2-eq/kWh for China and Europe. Taking an average of 80 kg CO_2-eq/kWh, based on 10 TWh production volume in 2040, to produce such a number of batteries will generate approximately 0.8 billion tons of CO_2-eq, which accounts for 4.7% of global total annual emissions if it follows the Paris Agreement decarbonization route, making the battery sector an important emission source (Figure 5.1).

5.1.2 Resource Depletion and Unsustainable Mining

The growing demand for batteries has led to the overexploitation of critical raw materials, which has resulted in severe price fluctuations in 2022–2024, e.g., the price of lithium carbonate rocketed to over 600,000 CNY/Mt in 2022 and fell back to below 100,000 CNY/Mt in 2024. Lithium, cobalt, and nickel are finite resources, and their extraction often involves environmentally destructive practices, including

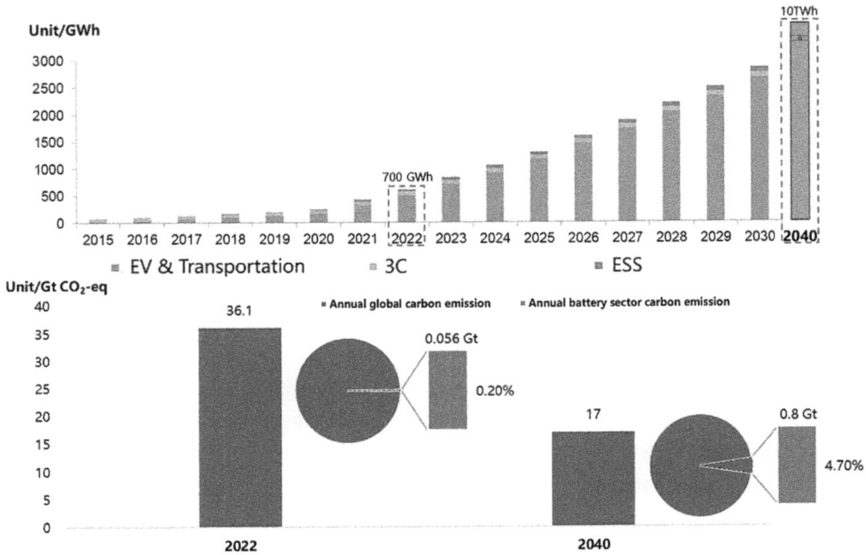

Figure 5.1 Forecast of carbon emissions of the battery industry and its contribution.

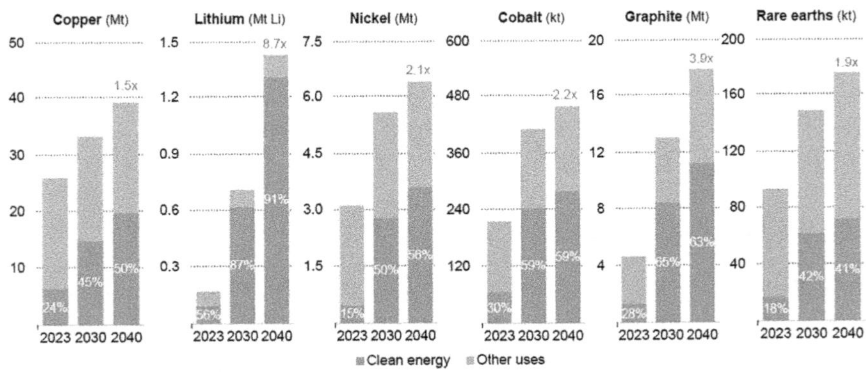

Figure 5.2 Global critical minerals demand in the net-zero emission (NZE) scenario.

water pollution and soil degradation. Additionally, the lack of traceability in supply chains makes it difficult to ensure ethical and sustainable sourcing, exacerbating issues like child labor and conflict minerals. There is still a big room to grow; as predicted by IEA [2], the demand for lithium, nickel, and cobalt will grow 8.7, 2.1, and 2.2 times in 2040 compared to the 2023 level, as shown in Figure 5.2.

5.1.3 Inadequate End-of-Life Management

Current lithium battery recycling rates have a big room to improve, according to Hans Eric Melin. By tracking the flows of EoL battery materials and manufacturing scrap from 2023, they estimated that 59% of the batteries that were available for recycling were recycled, leaving the rest of the 41% untracked [3], thus presenting a significant obstacle to the sustainable development of the battery industry.

This dismal figure is a stark contrast to the increasing demand for these critical raw materials, highlighting the urgent need for more effective recycling strategies.

The low recycling rates can be attributed to several factors. Technologically, current battery recycling processes are complex and costly. Different battery chemistries, such as lithium-NCM, LFP, lithium-manganese-oxide (LMO), and lithium-cobalt-oxide (LCO), require different recycling techniques. Traditional recycling methods, such as pyrometallurgy (high-temperature melting) and hydrometallurgy (chemical leaching), have limitations in balancing the metal recovery efficiency and environmental impact. Pyrometallurgical processes consume large amounts of energy and can release greenhouse gases, while hydrometallurgical processes generate large volumes of chemical waste.

Economically, the current market conditions do not provide sufficient incentives for widespread battery recycling. The cost of collecting, transporting, and recycling batteries often exceeds the value of the recovered materials, especially when the prices of virgin raw materials are relatively low. In 2024, when the price of lithium carbonate dropped significantly, the economic viability of recycling lithium from EoL batteries was severely affected, especially for LFP and LMO battery chemistries. This has led many recycling companies to operate at a loss or struggle to stay in business, hindering the development of the battery recycling industry.

In addition, there is a lack of public awareness and proper collection infrastructure for EoL batteries. Many consumers are still unaware of the environmental hazards of improper battery disposal or the availability of recycling options. In some regions, there are simply not enough collection points for consumers to drop off their used batteries, making it more convenient for them to discard the batteries along with regular waste.

The improper management of EoL batteries not only contributes to the growing electronic waste (e-waste) problem but also undermines the efforts to achieve a circular economy in the battery industry. Without significant improvements in battery recycling rates and EoL management, the sustainable development of the clean energy transition, which relies heavily on the widespread use of batteries, will be severely compromised.

5.2 Mitigating GHG Emissions Through Supply Chain Transparency

5.2.1 Raw Material Extraction/Processing and Cell Production

Battery Passports serve as a digital ledger that records every stage of a battery's lifecycle, from the extraction of raw materials to the final disposal. This transparency is crucial in identifying and reducing GHG emissions. Documenting raw material extraction sites in the Battery Passport allows companies to closely monitor the energy-intensive processes involved. The GHG emissions for the same product produced from different mine sources and processes can be dramatically different. Liu et al. have studied and compared the carbon footprint of lithium carbonate produced from different lithium resources, i.e., brine, spodumene, and lepidolite [4], as illustrated in Figure 5.3. Their study presented a wide range of product carbon footprints, 6.3–36.8 kg CO_2-eq/kg.

The study from Minviro [5] found similar patterns for nickel, whose carbon footprint can vary from 1 to 18 kg CO_2-eq/kg when produced from nickel sulfide and nickel laterite using different technologies (Figure 5.4).

This visibility can create an incentive for mining companies to switch to cleaner energy sources, such as electric-powered vehicles or renewable energy-powered machinery.

Moreover, the synthesis of cathode and anode active materials is another significant source of emissions. By tracking the manufacturing processes of these materials in the Battery Passport, companies can analyze the energy consumption patterns. They can then invest in research and development to find more energy-efficient production methods. For instance, researchers are exploring the use of microwave-assisted synthesis techniques for cathode materials, which can potentially reduce the energy required for the synthesis process and, consequently, lower GHG emissions [6]. The ability to trace these processes through the Battery Passport allows for continuous improvement and optimization of the manufacturing steps to minimize environmental impact.

In the battery cell production stage, the majority of GHG emissions result from the energy consumption, as illustrated in Figure 5.5 [1]. If the passport can record details such as the type and amount of energy used in the factory, it can push for the improvement of the production equipment efficiency and the shift to greener energy sources. This level of transparency, enabled by Battery Passports, empowers companies to make informed decisions that directly contribute to reducing GHG emissions throughout the battery production process.

5.2.2 Optimizing the Battery Lifecycle

Electric vehicle batteries, predominantly lithium-ion, are engineered for high performance but are subject to gradual degradation over time, which manifests as a slow reduction in their ability to hold an electrical charge. This degradation process is influenced by a complex interplay of operational and environmental factors, including specific charging behaviors, the magnitude of charging current, prolonged temperature exposure, the duration of use, and the cumulative number of charge and discharge cycles [7]. For instance, the frequent reliance on direct current (DC) fast chargers, while convenient, has been shown to accelerate battery aging compared to the less stressful, slower charging methods, such as overnight charging at home. Similarly, consistently charging a battery to its full 100% capacity or allowing its charge level to drop excessively low (typically below 20%) can induce significant stress on the internal battery cells, leading to accelerated wear [8]. Furthermore, extreme ambient temperatures, both excessively hot and severely cold, negatively impact battery longevity. Although modern EVs are equipped with sophisticated thermal management systems designed to maintain optimal battery temperatures, these systems consume battery power to operate. Therefore, in very

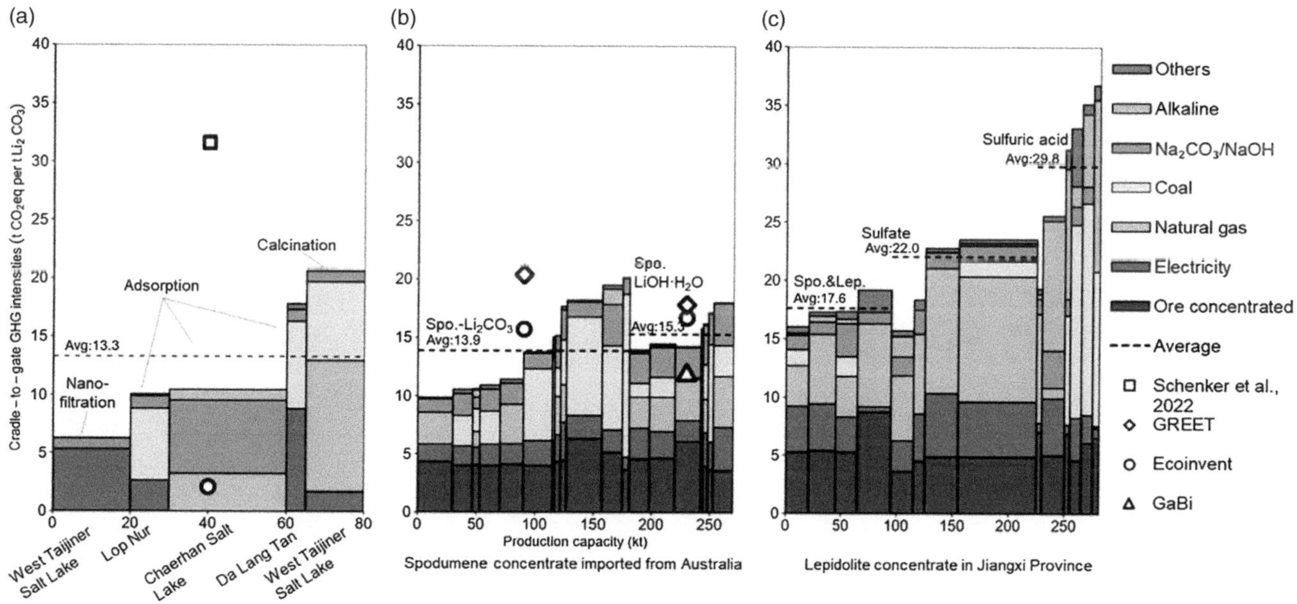

Figure 5.3 Plant-level carbon footprint for Li_2CO_3 produced from (a) brine, (b) spodumene, and (c) lepidolite.

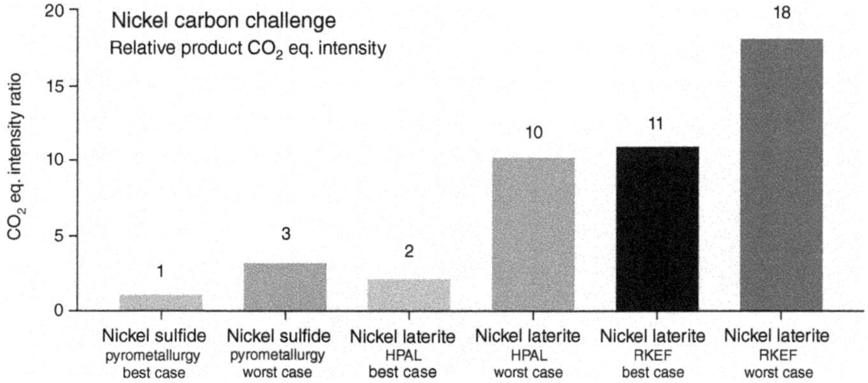

Figure 5.4 Carbon footprint of nickel sulfate hexahydrate production routes per technology.

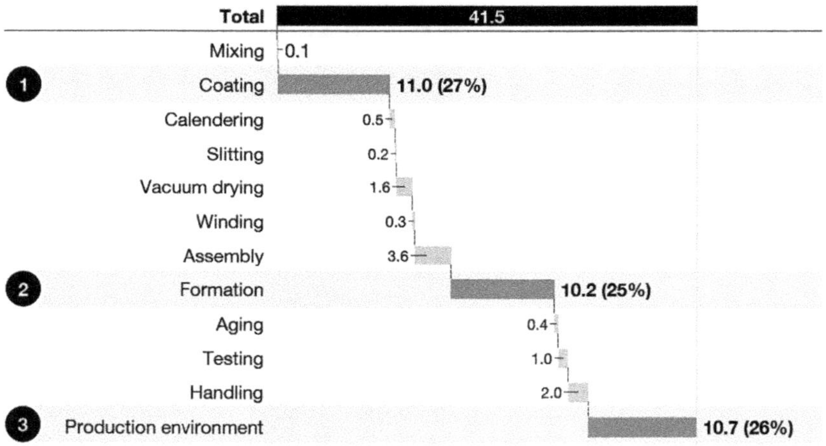

Figure 5.5 Energy consumption in lithium-ion cell production (kWh per kWh).

hot or very cold conditions, it is often advisable to keep the vehicle plugged in when parked, allowing the thermal management system to draw power from the grid rather than the battery itself. Parking in shaded areas in hot climates or in a garage in extreme cold can also contribute to maintaining a moderate temperature range, which is crucial for preserving battery health.

The interplay of user behavior, such as specific charging habits and driving patterns, and environmental factors, particularly temperature, is a primary driver of battery degradation. This degradation can often be more significant than variations attributed solely to inherent manufacturing differences. This observation highlights the critical need for dynamic data collection during the actual use phase of the battery. Static manufacturing data, while providing initial specifications, cannot fully capture the real-world stresses and conditions a battery experiences. Therefore, continuous, real-time monitoring of operational parameters is essential to accurately assess, predict, and ultimately mitigate battery degradation, enabling more effective lifecycle management strategies.

Digital Battery Passports are instrumental in extending the operational lifespan of EV batteries by providing precise tracking of their performance and health throughout their active use phase [9]. These passports collect dynamic data on critical operational parameters, including detailed charging patterns, temperature exposure profiles, discharge rates, real-time state of health (SoH) metrics, and projections of remaining useful lifetime [10].

The analysis of this rich dataset can reveal specific usage patterns that contribute to premature degradation. For example, if the data indicates that a vehicle owner frequently utilizes fast charging, the DBP (Digital Battery Passport) -enabled system can provide targeted recommendations for more optimal charging habits, such as prioritizing slower, overnight charging sessions more often. Beyond generic advice, advanced artificial intelligence (AI)-powered systems can even customize charging and discharging patterns based on an individual's unique driving habits, planned routes, and prevailing local climate conditions. This level of personalized optimization directly contributes to improving the vehicle's range and significantly extending the battery's lifespan [11].

This data-driven approach facilitates proactive maintenance interventions, thereby reducing the likelihood of unexpected battery failures and consequently extending their overall lifespan [10]. Furthermore, the DBP enhances transparency and visibility into the actual condition of EV batteries for various stakeholders, including consumers, OEMs, dealerships, and insurance providers. This empowers consumers to actively maintain their battery's health and gain a clearer understanding of their vehicle's long-term value [12].

The ability of Battery Passports to provide personalized recommendations based on real-time usage data transforms battery management from a generic, reactive process into a highly individualized, proactive optimization strategy. This represents a significant evolution beyond simple data logging, moving toward actionable intelligence. When a DBP collects granular data on an individual battery's usage, and AI processes this data to identify specific degradation patterns and recommend tailored optimizations (e.g., suggesting slower charging based on a user's driving profile), it creates a powerful opportunity. This capability can foster deeper consumer engagement, as users receive direct, relevant advice for preserving their assets. Moreover, it opens avenues for OEMs to introduce new service-based revenue models, such as "Battery Health Optimization Subscriptions," where continuous monitoring and personalized guidance are offered as a premium service. This strategic shift allows OEMs to diversify their business portfolios and build stronger, long-term relationships with their customers, moving beyond a one-time vehicle sale to a continuous value proposition.

AI and machine learning (ML) are pivotal in significantly enhancing the capabilities of Battery Passports, particularly in extending the longevity of EV batteries. These technologies enable a more sophisticated and predictive approach to battery management.

Real-time health monitoring and predictive analytics are core applications. By integrating AI with Internet of Things (IoT) sensors embedded directly within batteries, continuous monitoring of critical parameters, such as charge cycles, internal temperature, and voltage levels, becomes feasible. AI algorithms are then employed to process this vast influx of data, enabling accurate assessment of battery performance, early detection of degradation signs, and precise prediction of future health trends. This proactive capability allows for timely maintenance interventions, drastically reducing the likelihood of unexpected battery failures and consequently extending their operational lifespan.

Furthermore, AI-driven battery management systems (BMS) can determine optimal charging protocols. These systems analyze factors such as the battery's age, current temperature, and historical usage patterns to dynamically adjust charging strategies. This intelligent management reduces stress on the battery cells, minimizes degradation, and ultimately extends the battery's useful life [13]. This level of personalization, informed by AI, can significantly improve both the vehicle's driving range and the overall battery lifespan.

The convergence of AI, IoT, and advanced BMS within the Digital Battery Passport framework represents a profound shift from reactive battery management to a highly intelligent, self-optimizing system. This integrated approach is not merely about extending battery life; it fundamentally redefines the relationship between the vehicle, its battery, and the user. The continuous flow of data from IoT sensors, processed by AI algorithms, allows the BMS to learn from usage patterns and environmental conditions in real time. This learning capability enables the system to adapt dynamically, predicting potential issues and adjusting operations proactively. This transformation paves the way for "smart" battery ecosystems that continuously learn, adapt, and optimize their performance, leading to unprecedented levels of efficiency, reliability, and user confidence in electric mobility.

5.2.3 Encouraging Sustainable Sourcing

Battery Passports play a pivotal role in promoting sustainable sourcing practices by enabling end-to-end traceability of raw materials, ensuring they are extracted and processed in an environmentally and socially responsible manner. Transparency into the origin of materials such as lithium, cobalt, nickel, and graphite allows companies to identify suppliers that adhere to sustainable standards – such as low-carbon extraction methods, fair labor practices, and avoidance of conflict minerals – and incentivize others to adopt similar practices.

For instance, the inclusion of mineral origin data in Battery Passports can help mitigate environmental harm from artisanal mining, which often lacks proper regulation and contributes to deforestation, water pollution, and GHG emissions. By publicly disclosing such data, Battery Passports create market pressure for suppliers to adopt cleaner technologies, such as renewable energy-powered processing plants or water-recycling systems in lithium extraction.

Moreover, digital passports facilitate compliance with international sustainability standards, such as the EU's Corporate Sustainability Due Diligence Directive

(CSDDD) or the OECD's Due Diligence Guidance for Responsible Supply Chains of Minerals. For example, tracking cobalt from Democratic Republic of Congo (DRC) mines through Battery Passport can ensure it is free from child labor and forced labor, aligning with ethical sourcing requirements. This level of accountability not only reduces reputational risks for manufacturers but also meets growing consumer demand for "green" products – surveys indicate that over 60% of global consumers are willing to pay a premium for sustainably sourced electronics [14]. By embedding sustainable sourcing criteria into Battery Passport frameworks, the industry can drive systemic change. The transparency enabled by passports creates a virtuous cycle where sustainable practices are rewarded through preferential sourcing agreements, ultimately reducing the environmental and social toll of battery production.

5.3 The Role of Battery Passports in Achieving Closed-Loop Recycling

5.3.1 Improving Collection and Sorting

One of the major hurdles in battery recycling is the efficient collection and sorting of EoL batteries. Battery Passports can revolutionize this process. Since the passport contains detailed information about the battery's type, chemistry, and usage history, it becomes easier to identify and collect batteries at the end of their life. For example, in a large-scale EV fleet, the Battery Passports of each vehicle's battery can be used to track when the battery is approaching the end of its useful life. Fleet managers can then proactively arrange for the collection of these batteries, ensuring that they are not left in vehicles or disposed of improperly.

In addition, the information in the Battery Passport helps in the sorting of batteries. Different battery chemistries require different recycling processes. Lithium-ion batteries with different cathode chemistries, such as NCM and LFP, have different compositions and recycling requirements. By using the data from the Battery Passport, recycling facilities can quickly and accurately sort batteries according to their chemistry, ensuring that each type of battery is processed using the most appropriate and efficient recycling method. This not only improves the quality of the recycled materials but also increases the overall efficiency of the recycling process, reducing the energy consumption and emissions associated with improper sorting and processing.

5.3.2 Enabling Second-Life Applications

Upon reaching the end of their first-life utility in demanding applications such as EVs or primary energy storage systems, batteries typically retain a substantial portion of their original capacity. For EV batteries, this often amounts to 70–80% of their initial capacity. This significant residual capacity renders them highly suitable for less demanding "second-life" applications, where their performance characteristics remain adequate.

From an economic perspective, repurposing these batteries captures the inherent value of their remaining useful life. This generates additional revenue streams for original battery owners, such as OEMs who may retain ownership through leasing programs, and simultaneously reduces costs for entities utilizing them in second-life applications. Furthermore, by extending the operational life of existing batteries, the need for new battery production is delayed, which represents a significant cost factor in the overall battery value chain. The economic feasibility of second-life batteries is intricately tied to their acquisition price and their assessed technical properties, making a clear valuation critical.

Environmentally, the reuse of batteries offers profound benefits. It directly extends the battery lifecycle, leading to a significant reduction in the demand for newly manufactured batteries and the associated GHG emissions from their production. By embracing reuse, valuable materials are kept within the economic system for a longer duration, directly contributing to the principles of a circular economy and minimizing waste.

The economic viability of second-life batteries is intrinsically linked to their environmental benefits. The ability to extract residual value from used batteries creates a powerful incentive for businesses to adopt circular practices. When companies realize that batteries, even after their primary use, retain significant capacity that can be monetized in less demanding applications, it encourages investment in repurposing infrastructure and processes. This financial incentive drives a strategic shift in the industry from a purely linear "take-make-dispose" model to one that prioritizes resource efficiency and extended product utility. This transition not only generates new revenue streams and reduces costs but also directly contributes to environmental sustainability by decreasing the demand for virgin materials and the energy-intensive processes of new battery manufacturing. This creates a virtuous cycle where economic gain directly propels environmental preservation.

The rigorous assessment of a battery's fitness for repurposing is a critical step in enabling its second life. This process involves a detailed evaluation of its technical properties after its initial use, including its remaining capacity, operational temperature history, cumulative cycle life, and internal resistance. This comprehensive assessment ensures that each battery is optimally matched with the most appropriate second-life application, thereby maximizing its utility and efficiency in its new role. This often involves a grading system, where batteries are categorized based on their SoH, for instance, from Grade A (above 90% SoH) to Grade D (50–70% SoH), to channel them into suitable reuse or repurposing pathways.

A significant challenge in this assessment has historically been the availability of traceable information regarding the battery's first-life history and its SoH. This crucial data is typically held by the original vehicle OEM, and its accessibility is vital for repurposers to make informed and reliable decisions about a battery's suitability. To standardize this process, frameworks and standards such as UL 1974 provide guidelines for evaluating batteries for repurposed use, encompassing considerations for cell evaluation and various usage scenarios.

The historical lack of transparent, standardized data on a battery's first-life usage and degradation has been a significant impediment to scaling second-life markets. Repurposers often faced substantial uncertainty and incurred high costs in individually assessing the condition of used batteries due to fragmented or unavailable historical data. The Digital Battery Passport directly addresses this information asymmetry by providing verifiable, comprehensive data on a battery's entire lifecycle, including its performance history, SoH, and any recorded incidents. By making this critical information readily accessible and trustworthy, the DBP de-risks investments in second-life applications and significantly streamlines the assessment and reclassification processes (the study from Battery Pass suggests the testing-related cost reduction could reach 2–10% [15]), thereby accelerating the growth and efficiency of the secondary battery market (Figure 5.6).

5.3.3 Facilitating Efficient Recycling

Efficient battery recycling constitutes a pivotal element in the realization of closed-loop systems, as it dictates the quality and quantity of recoverable materials for subsequent reuse. Conventional recycling methodologies frequently grapple with inefficiencies stemming from inadequate material traceability, suboptimal sorting techniques, and limited insight into a battery's internal condition, thereby resulting in substandard recovery rates and heightened environmental repercussions. Battery Passports emerge as a revolutionary instrument in this domain, harnessing digital data to streamline every phase of the recycling process – from disassembly to material recovery – while simultaneously enhancing both economic viability and ecological sustainability.

5.3.3.1 Material Traceability and Dismantling Process Optimization

Central to efficient recycling lies the capability to trace a battery's composition and structural history. Battery Passports store granular data pertaining to a battery's design, encompassing its cell chemistry (e.g., NCM, LFP, or solid-state), electrode materials, electrolyte composition, and thermal management systems. This information holds immense value for recycling facilities, as it enables them to preplan disassembly strategies tailored to the battery's specific architectural configuration. For instance, a battery featuring a modular design (a common characteristic in modern EVs) can be disassembled with greater efficiency when recyclers are apprised of the precise configuration and fastening mechanisms of its cells and packs, thereby minimizing damage to recoverable components.

Furthermore, the passport's documentation of a battery's usage history – encompassing charging cycles, temperature exposure, and mechanical stress – affords insights into its degradation patterns. Batteries that have undergone severe thermal stress, for example, may necessitate specialized handling to preclude hazardous reactions during the disassembly process. By flagging such characteristics, the passport empowers recyclers to prioritize safety protocols and allocate

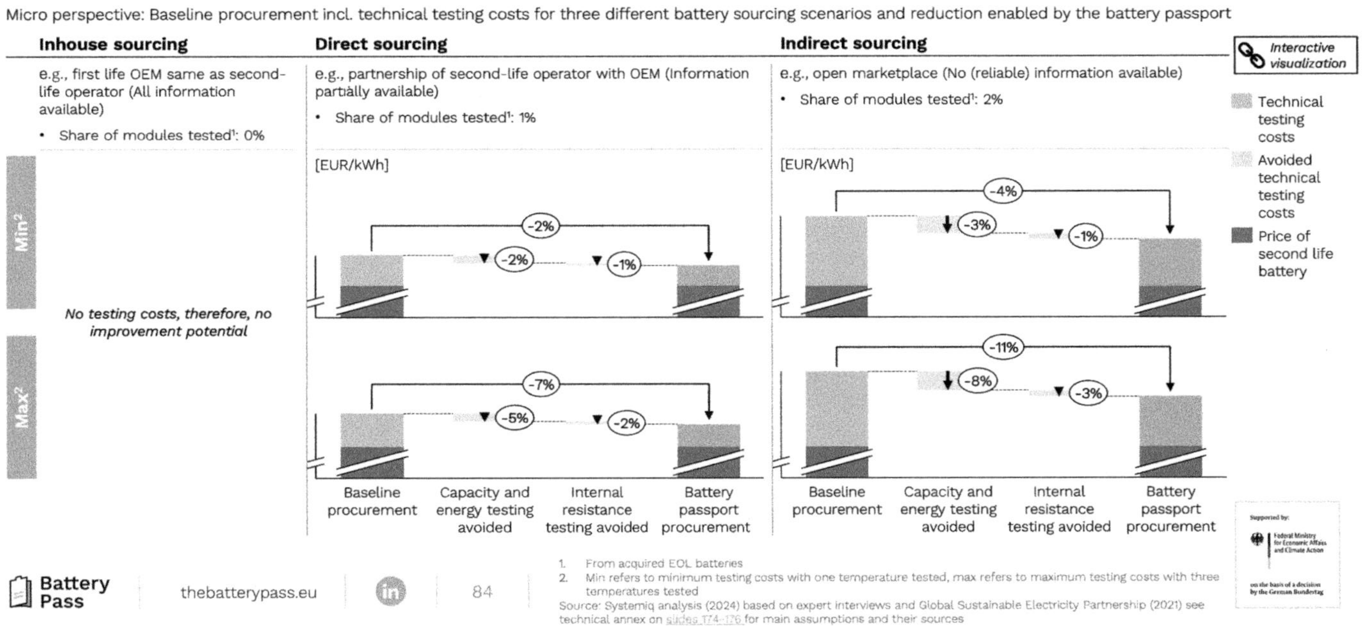

Figure 5.6 Technical testing costs for three different battery sourcing scenarios and reduction enabled by the Battery Passport. *Source:* Battery Pass [16] / Systemiq / CC BY 4.0.

resources strategically, thereby mitigating risks of fire, electrolyte leakage, or material contamination.

5.3.3.2 Precision Sorting for Chemical Recycling Applications

As elaborated in Section 5.3.1, divergent battery chemistries necessitate distinct recycling approaches. Battery Passports elevate this process by furnishing real time, verifiable data on a battery's chemical composition, thereby enabling automated sorting systems to categorize units with unparalleled precision. Consider, for example, a recycling facility processing a mixture of NCM and LFP batteries. Traditional manual sorting is susceptible to errors and inherently time-consuming, whereas passport data enables robotic arms to swiftly direct each battery to the appropriate processing line. This reduces cross-contamination, which can compromise the quality of recycled materials and complicate refining procedures. Through optimization of the sorting process, Battery Passports enhance the purity of recovered materials, thereby increasing their value for reuse in new batteries or other industrial applications. Based on the study from Battery Pass, it can lower the cost for pretreatment and recycling by 10–20% [15] (Figure 5.7).

5.3.3.3 Economic Incentives for Eco-Design

Furthermore, the passport facilitates the emergence of "closed-loop" partnerships between manufacturers and recyclers. OEMs or battery producers can utilize passport data to design batteries with recycling in mind (e.g., by employing fewer adhesives or standardized cell formats), while recyclers can guarantee the quality of recovered materials to manufacturers. This symbiotic relationship fosters a circular economy in which material value is retained within the supply chain, reducing reliance on costly raw material imports and promoting regional self-sufficiency in critical resources such as lithium and cobalt.

5.3.4 Promoting the Use of Recycled Materials

Battery Passports can also play a significant role in promoting the use of recycled materials in the production of new batteries. By clearly documenting the origin and quality of recycled materials used in a new battery, the passport can increase consumer and industry confidence in recycled-based batteries. For example, if a battery is made with a certain percentage of recycled lithium, cobalt, or nickel, this information can be prominently displayed in the Battery Passport. This not only showcases the environmental benefits of using recycled materials but also differentiates the product in the market.

Moreover, the Battery Passport can create a feedback loop between the recycling and production sectors. As more information about the performance and quality of batteries made with recycled materials is collected through the passport, manufacturers can use this data to continuously improve their recycling and production processes. If a particular batch of recycled materials results in batteries with lower performance, the data can be used to identify the root cause, whether it is a problem with the recycling process or the integration of recycled materials into the new

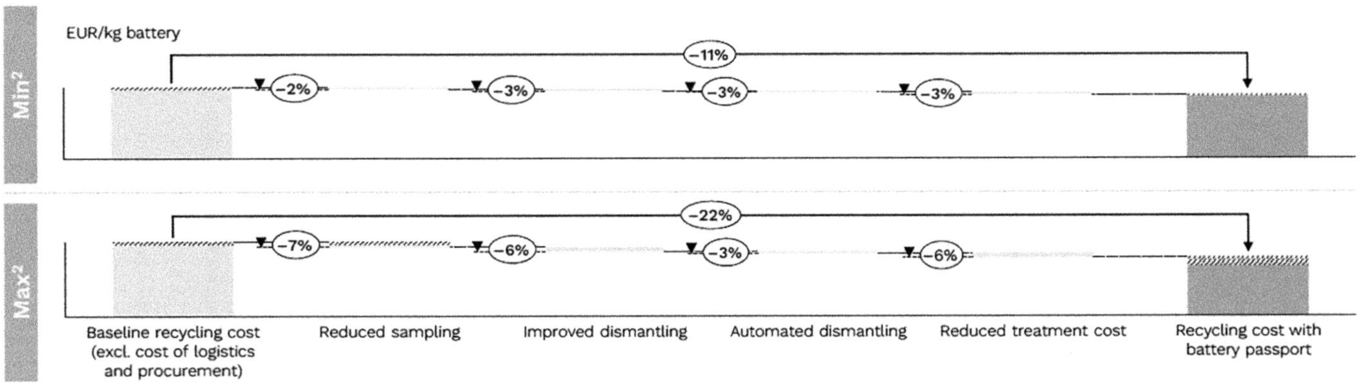

Figure 5.7 Recycling cost with the Battery Passport. *Source:* Battery Pass [16] / Systemiq / CC BY 4.0.

battery production. This iterative process of improvement encourages the increased use of recycled materials in battery production, bringing the industry closer to achieving a closed-loop recycling system where the demand for virgin raw materials is significantly reduced, and the environmental impact of battery production and disposal is minimized.

5.4 Conclusion

The global transition to electrification and renewable energy hinges on the sustainable management of lithium-ion batteries, yet their lifecycle currently poses significant environmental challenges, including high GHG emissions, critical raw material depletion, and inadequate recycling practices. This chapter has demonstrated that Battery Passports, as digital tools enabling end-to-end lifecycle tracking, are indispensable for addressing these issues and fostering a circular economy in the battery industry:

1) **Mitigating GHG Emissions Through Transparency:** By documenting every stage from raw material extraction to cell production, Battery Passports expose carbon-intensive processes. For example, they reveal stark differences in carbon footprints between lithium carbonate sourced from brine (6.3 kg CO_2-eq/kg) vs. lepidolite (36.8 kg CO_2-eq/kg) and nickel production routes (1–18 kg CO_2-eq/kg). This visibility incentivizes shifts to cleaner energy sources in mining and manufacturing, such as renewable-powered machinery, and drives innovation in energy-efficient synthesis techniques for cathode materials. Additionally, tracking real-time energy use in production facilities enables targeted improvements in equipment efficiency and green energy adoption, directly reducing emissions.
2) **Enabling Closed-Loop Recycling:** Battery Passports revolutionize EoL management by optimizing collection, sorting, and recycling. Detailed data on battery chemistry (e.g., NCM vs. LFP) streamlines sorting for specialized recycling processes, reducing pretreatment costs by 10–20%. They also facilitate second-life applications: EV batteries with 70–80% residual capacity can be repurposed for energy storage, delaying the need for new production and cutting GHG emissions. By standardizing battery health assessments (e.g., SoH grading) and reducing testing costs by 2–10%, passports accelerate the growth of secondary markets. Furthermore, traceable material origins in passports promote the use of recycled lithium, cobalt, and nickel in new batteries, enhancing consumer trust and closing the material loop.
3) **Driving Sustainable Sourcing and Accountability:** The passports' traceability features ensure ethical and low-carbon sourcing of critical minerals. For instance, tracking cobalt from the DRC through passports helps eliminate child labor and conflict minerals, aligning with EU regulations like CSDDD. Public disclosure of sourcing data creates market pressure for sustainable practices, such as renewable energy in lithium extraction, while meeting consumer demand for "green" products.

While Battery Passports offer transformative potential, their effectiveness depends on global standardization, interoperability across supply chains, and robust policy frameworks. Harmonizing data formats, integrating AI-driven analytics for predictive maintenance, and fostering collaboration between OEMs, recyclers, and policymakers will be crucial.

References

1 Berger, R. (2024). Battery monitor 2024/2025, the value chain between economy and ecology. https://www.rolandberger.com/en/Insights/Publications/Battery-Monitor-2024-2025-A-turbulent-year-and-outlook-for-value-chain-players.html (accessed 14 May 2025).

2 International Energy Agency (2024). Global critical minerals outlook 2024. https://www.iea.org/reports/global-critical-minerals-outlook-2024 (accessed 2 August 2025).

3 Melin, H.E. (2024). The 5% rate and other untruths about battery recycling. LinkedIn (Pulse) [Internet]. Jan. 17. https://www.linkedin.com/pulse/5-rate-other-untruths-battery-recycling-hans-eric-melin-m5lde/ (accessed 27 March 2025).

4 Liu, M., Wang, F., Zhang, S. et al. (2025). Carbon footprint of battery-grade lithium chemicals in China. *ACS Sustainable Chemistry & Engineering*. https://pubs.acs.org/doi/10.1021/acssuschemeng.4c08394.

5 Minviro (2023). Nickel's carbon challenge, understanding the relationship between nickel source and carbon intensity. https://26877240.fs1.hubspotusercontent-eu1.net/hubfs/26877240/White%20papers/Minviro_Nickel_WhitePaper_Jun2023.pdf?__hstc=74532941.828b52090c030b362c739469379d3d92.1749346319545.1749346319545.1749346319545.1&__hssc=74532941.5.1749346319545&__hsfp=1541238644 (accessed 9 July 2025).

6 Styuf, E.A., Othman, M., Savina, A.A. et al. (2025). The microwave-assisted spray drying synthesis: application to the $LiFePO_4$/C cathode material for Li-ion batteries. *Journal of Power Sources*. https://www.sciencedirect.com/science/article/abs/pii/S0378775325007876.

7 Terkes, M., Demirci, A., Gokalp, E. et al. (2024). Battery Passport for second-life batteries: potential applications and challenges. *IEEE Access* 128424–128467. https://eprints.whiterose.ac.uk/id/eprint/222123/1/Battery_Passport_for_Second-Life_Batteries_Potential_Applications_and_Challenges.pdf.

8 Milchtein, C.M. (2021). How to extend the life of electric vehicle batteries. https://mwg.aaa.com/via/car/how-extend-life-electric-vehicle-batteries (accessed 18 April 2025).

9 DigiProd Pass. (2025). EU digital Battery Passport: the complete guide. https://digiprodpass.com/blogs/digital-battery-passport-what-it-is-who-must-comply-and-when/ (accessed 6 June 2025).

10 CIRCULARISE. (2025). EU Battery Passport regulation requirements. https://www.circularise.com/blogs/eu-battery-passport-regulation-requirements (accessed 21 August 2025).

11 The Times of India. (2025). AI-powered Battery Passport: shaping the future of battery management and innovation. https://timesofindia.indiatimes.com/auto/electric-cars/ai-powered-battery-passport-shaping-the-future-of-battery-management-and-innovation/articleshow/118786974.cms (accessed 30 May 2025).

12 PERFICIENT. (2025). The new standard for electric vehicle health and value. https://www.perficient.com/insights/strategic-positions/digital-technologies/battery-passport (accessed 11 March 2025).

13 ELECTRA. (2024). From passive to adaptive: the rise of AI-driven battery management systems. https://www.electravehicles.com/from-passive-to-adaptive-the-rise-of-ai-driven-battery-management-systems/ (accessed 25 July 2025).

14 Dwf. (2024). Consumer trends 2024: transparency as a responsible business force for good. https://dwfgroup.com/en/news-and-insights/insights/2024/1/ct24-transparency-as-a-responsible-business-force-for-good (accessed 3 April 2025).

15 Herrmann, S., Niemann, N., Vahle, T., et al. (2024). Unlocking the value of the EU Battery Passport: an exploratory assessment of economic, environmental and social benefits. Value Assessment of the Battery Passport, Version 1.0 [Internet]. Battery Pass consortium; November 2024. https://thebatterypass.eu/assets/images/value-assessment/pdf/2024_BatteryPassport_Value_Assessment.pdf (accessed 19 June 2025).

6

Global Battery Passports Case Study: A Comprehensive Analysis of Initiatives

6.1 GBA Battery Passport Initiative

6.1.1 The Vision of GBA

Established in 2017 through a collaborative effort between the World Economic Forum and multiple stakeholders, the Global Battery Alliance (GBA) is dedicated to establishing a circular, responsible, and equitable battery value chain by 2030. As of 2024, GBA has expanded its membership to 175 entities, encompassing industry leaders such as CATL and Volkswagen, along with esteemed institutions like the United Nations Environment Programme and the International Labour Organization.

The EU Battery Regulation (2023/1542) explicitly mandates the compulsory implementation of a "digital passport" for batteries beginning in 2027. GBA's framework design not only directly aligns with this legislative requirement but also seeks to extend its applicability on a global scale. This initiative addresses pressing challenges within the battery supply chain, including issues related to child labor and ecological damage. The GBA Battery Passport leverages digital tools to track data throughout the entire lifecycle of batteries, with the aim of ensuring fairness in the clean energy transition and preventing the overexploitation of vulnerable regions during the extraction and processing of resources.

In pursuit of its 2030 Vision, GBA has identified three significant challenges facing the battery value chain:

1) **Sustainability of Raw Material Extraction:** Addressing ecological harm, child labor concerns, and disputes over community rights that arise during the mining of critical minerals such as cobalt and lithium.
2) **Low-carbon Production and Utilization:** Reducing carbon emissions in battery manufacturing through technological innovation and promoting the green adoption of electric vehicles (EVs) and energy storage systems.
3) **Development of a Circular Economy System:** Establishing an efficient waste battery recycling network to minimize resource waste.

6.1.2 Sustainability Indicator Framework and Scoring Logic

To achieve its objectives, GBA has established a robust framework of guiding principles encompassing environmental, social, and governance (ESG) dimensions and launched the Battery Passport piloting program. This initiative aims to document comprehensive lifecycle data of batteries, including raw material origins, production processes, carbon footprints, usage patterns, and recycling trajectories, with the overarching goal of enhancing supply chain transparency and strengthening compliance oversight. For instance, GBA has collaborated with automotive manufacturers, battery producers, and other industry stakeholders on pilot projects to integrate the Battery Passport into EV supply chains, thereby advancing sustainable development across the sector (Figure 6.1).

Cross-Cutting

Due Diligence
1. Presence and quality of environmental & human rights due diligence and risk management systems

Environmental

Energy and GHG
2. EGE emissions
3. Energy efficiency

Environmental Degradation
4. Pollution (air including GHG emissions, water, soil, hazardous substances, noise and vibration, plant safety
5. Biodiversity loss
6. Water management (usage, recycling, depletion)
7. Waste management (generation, recycling/reuse)

Circularity
8. Material consumption/usage
9. Circular design

Social & Economic

Human Rights
10. Child labor
11. Forced labor

Workers' Rights
12. Freedom of association and collective bargaining
13. Worker health and safety

Community Impacts and Rights
14. Respect for Indigenous Peoples' rights
15. Community life
16. Diversity

Local Economy
17. Contribution to local economic development (payments to governments and local supplies and employment
18. Engagement with artisanal and small-scale miners

Product Cost
19. Total cost of ownershipp

Governance

Compliance & Good Governance
20. Product quality and safety
21. Data security and privacy
22. Occurance of corruption and bribery

Legend:
- ESG issues corresponding to the risk categories of the EU Battery Regulation
- Additional salient ESG issues

Figure 6.1 GBA Battery Passport indicator framework. *Source:* Global Battery Alliance [1] / Global Battery Alliance.

As a core instrument for GBA to drive forward the sustainable battery value chain, its technical architecture and functional design are anchored in four foundational pillars.

6.1.2.1 Sustainability Indicator System

GBA's ESG framework is structured across three dimensions:

- **Environmental Dimension:** Focuses on energy efficiency, greenhouse gas (GHG) emission management, pollution control, water resource stewardship, and circular economy design. The *Greenhouse Gas Rulebook* provides standardized methodologies for carbon footprint calculations, aligning with and exceeding regulatory requirements.
- **Social and Economic Dimension:** Encompasses human rights protections (e.g., prohibition of child labor and forced labor), labor rights (occupational health and safety), community rights (indigenous land and cultural preservation), and local economic contributions (employment generation and support for artisanal mining).
- **Governance Dimension:** Emphasizes compliance management (product safety and quality assurance), data privacy safeguards, and anti-corruption practices.

6.1.2.2 Three-Tier Scoring Mechanism

The scoring system adopts a progressive model:

- **Basic Level:** Requires enterprises to establish a minimum viable management system and publicly declare sustainability commitments.
- **Intermediate Level:** Mandates third-party audits to verify compliance with Battery Passport criteria, culminating in formal certification.
- **Benchmark Level:** Targets aspirational standards (e.g., zero child labor, 100% renewable energy adoption) and awards the highest scores for exceptional performance.

6.1.2.3 Distributed Ledger Technology (DLT) for Data Integrity

Blockchain and other DLT solutions are employed to ensure data authenticity, immutability, and security, aligning with the stringent transparency requirements of regulations such as the EU Green Claims Directive. This technology underpins the traceability and credibility of sustainability data throughout the battery lifecycle.

6.1.2.4 Regulatory Compliance and Technical Integration

The Battery Passport supports enterprises in meeting legislative obligations, such as those outlined in the EU Battery Regulation (2023/1542), while encouraging practices that exceed minimum compliance thresholds. Technically, data collection is structured across three tiers:

- **Site Level:** Captures management systems of factories and mines.
- **Supply Chain Level:** Traces raw material provenance using advanced traceability technologies.
- **Battery Level:** Assigns unique identifiers (e.g., QR codes) to individual batteries for end-to-end tracking.

GBA adopts a collaborative approach, partnering with 11 technology providers to integrate sustainability indicators into existing digital platforms, rather than developing standalone solutions. This strategy fosters a cross-industry data ecosystem, enabling interoperability and scalable implementation of the Battery Passport across global supply chains.

6.1.3 2023 GBA Battery Passport POC (Proof of Concept) Pilot

The initial phase of the pilot program (2023), spearheaded by automotive manufacturers including Audi and Tesla, successfully achieved the first comprehensive integration and display of carbon footprint data, human rights due diligence information, and material traceability data. Spanning three subphases from November 2021 to December 2022, the pilot's core objectives were to authenticate data standardization protocols and assess cross-industrial chain collaboration frameworks. Notwithstanding these achievements, the current dataset remains significantly constrained, as it only encompasses partial raw materials such as lithium, cobalt, and nickel in its statistical scope.

6.1.3.1 Core Indicators and Data Attributes

GHG emission data emerged as a focal point of the pilot. Data analytics reveal that 59.6% of the emissions data originated from primary sources collected directly, while the remaining 40.4% was derived from literature-based estimations. Although the number of participating enterprises grew from 5 to 9, the traceability capabilities for critical components like positive electrode materials remained inadequate. An analysis of supply chain transparency indicated relatively comprehensive data coverage across mining, refining, and cell manufacturing stages, albeit with low overall interoperability scores. In terms of data validation, the project relied on third-party agencies for information collection but refrained from conducting in-depth audits of IT systems (Figure 6.2).

6.1.3.2 Key Insights and Areas for Improvement

While the pilot program validated the technical feasibility and practical utility of the Battery Passport, it also uncovered that approximately 12% of the mining areas under scrutiny posed child labor risks. This discovery has compelled enterprises to enhance their supply chain due diligence procedures to ensure greater compliance with ethical and sustainability standards.

6.1.4 2024 GBA Battery Passport MVP Pilot

The second-phase pilot in 2024 witnessed a substantial expansion in scale. Ten successful piloting consortia, among which are CATL, LG Energy Solution, and other enterprises representing 80% of the global EV battery production capacity, engaged in the testing of new rulebooks, such as those related to biodiversity loss

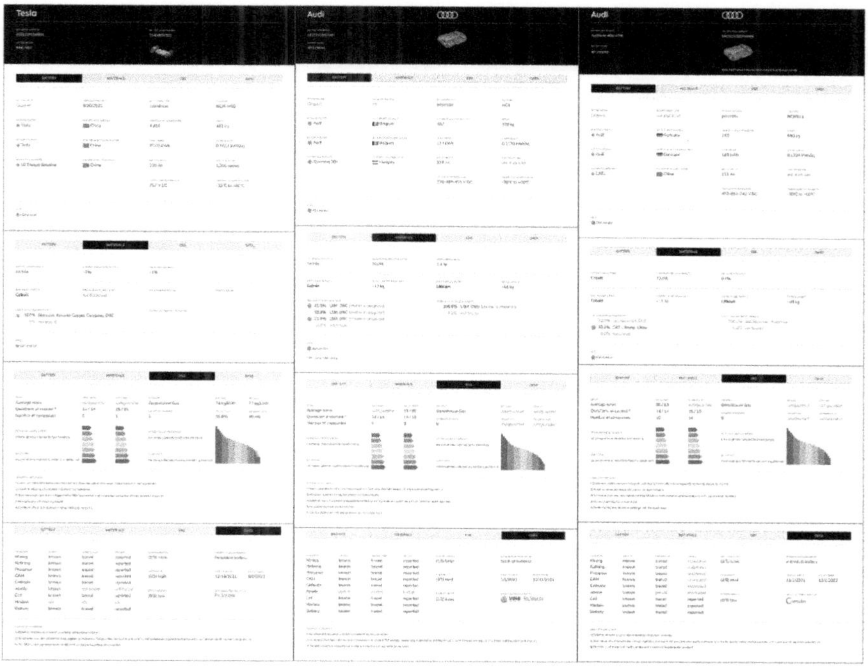

Figure 6.2 GBA Battery Passport pilot 2023 results. *Source:* Global Battery Alliance [2] / Global Battery Alliance.

and forced labor. Through a decentralized data exchange framework, this pilot project examined the sustainability and traceability of battery supply chains. The scope encompassed links including mineral mining, processing, manufacturing, and recycling, covering six materials (lithium, graphite, aluminum, cobalt, copper, and iron phosphate) from five continents [3].

The project validated GHG accounting methods (physical modeling and market coordination approaches) and achieved the first standardized carbon footprint scoring for each stage of the supply chain (divided into five clusters). However, due to vague data aggregation and boundary definitions (e.g., carbon footprint data for nickel and lithium not being subdivided by cluster), further optimization is required. The pilot found that complex third-party verification processes led to inefficiencies, and small and medium-sized enterprises faced high reporting costs. In the future, these issues will be addressed through simplifying verification rules, implementing a tiered reporting mechanism, and establishing a diversified verification ecosystem. Additionally, supply chain enterprises have high sensitivity to data sharing, necessitating a balance between transparency and privacy protection through decentralized data governance and digital signature technologies. Next steps include improving carbon footprint benchmark testing, refining the criteria for determining "not applicable" rules, and launching a beta pilot in 2025 to promote full compliance by 2027 (Figure 6.3).

Figure 6.3 GBA Battery Passport pilot 2024 results from CATL and EVE. *Source:* Global Battery Alliance [3] / Global Battery Alliance.

6.2 Battery Pass Initiative

6.2.1 Project Briefing

The Battery Pass project, a collaborative initiative in the EU led by Systemiq and colaunched by 11 industry-leading enterprises, research institutions, and technology service providers, aims to facilitate the digital implementation of the EU Battery Regulation. Funded by Germany's Federal Ministry for Economic Affairs and Climate Action (BMWK), the project is centered on establishing the technical framework and industry guidelines for the Battery Passport. This is intended to achieve standardized management and sharing of battery lifecycle data. Its core objectives are as follows:

- **Compliance Support:** Fulfilling the EU regulations' mandatory disclosure requirements regarding battery carbon footprint, supply chain due diligence, and recycling efficiency.
- **Value Chain Collaboration:** Enhancing resource efficiency in various links of battery design, use, recycling, etc., through data interconnectivity.
- **Circular Economy Enablement:** Prolonging the battery lifecycle, minimizing resource waste, and assisting Europe in achieving a green economic transition.

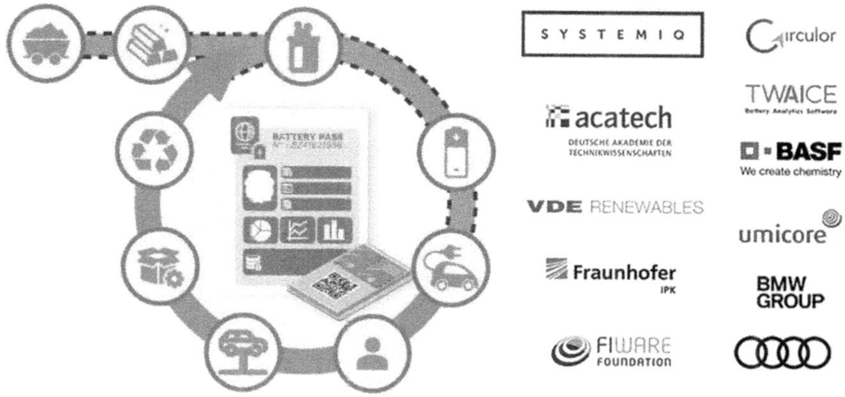

Figure 6.4 Member of Battery Pass project. *Sources:* (a–k) Battery Pass [4] / Battery Pass. Photo: Battery Pass [4] / Systemiq / CC BY 4.0.

The project's deliverables encompass technical architecture design, data standard formulation, software prototype development, and a value evaluation system. These provide a model reference for the digital governance of the global battery industry (Figure 6.4).

6.2.2 Battery Passport Architecture Design from Battery Pass

The Battery Pass Battery Passport system is designed with modular architecture, realizing full lifecycle data governance through standardized interfaces and distributed data management. Its core is made up of the EC Central Services and the Distributed Digital Product Passport (DPP) System Service Layer. Relying on a central registry as the data hub enables cross-platform data interoperability via API interfaces (Figure 6.5).

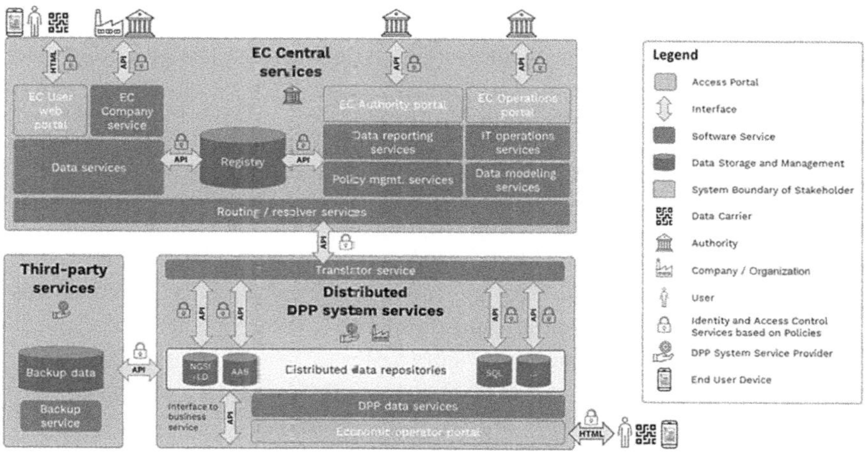

Figure 6.5 Battery Passport architecture illustrated by Battery Pass. *Source:* Battery Pass [5] / Systemiq / CC BY 4.0.

6.2.2.1 EC Central Services

- **User Interaction Layer**
  - EC User Web Portal: Serves end users.
  - EC Company Service: Provides enterprise-level services.
  - Both submit data to the central registry through APIs.
- **Regulatory Management Layer**
  - EC Authority Portal and EC Operations Portal carry out compliance audits and dynamic updates of Battery Passport information.
  - Equipped with modules for data reporting, policy management, and IT operational services.
- **Data Security Mechanism**
  - The central registry uses encrypted transmission (HTTPS) and identity authentication to ensure data security.

6.2.2.2 Distributed Digital Product Passport (DPP) System Service Layer

- **Data Conversion and Storage**
  - Integrates the Translator Service and Distributed Data Repositories.
  - Supports standardized conversion of heterogeneous data formats (such as Next Generation Service Interfaces – Linked Data [NGSI-LD], Asset Administration Shell [AAS]).
  - Enables efficient storage via database technologies like SQL.
- **Enterprise Interface**
  - The Economic Operator Portal acts as the terminal interface for enterprises to directly access data services.
  - Supported functions include battery status tracking, residual value assessment, and supply chain traceability.

6.2.2.3 System Supplementary Capabilities

- Integrates third-party backup services and data modeling services to ensure data redundancy and business continuity.

Through this layered architecture, it can achieve end-to-end data transparency throughout the battery's production, use, and recycling processes. It offers technical support for EU regulatory enforcement, circular economy optimization, and market value extraction.

6.2.3 Data Framework and Functions

The Battery Pass project has developed a dynamic digital platform that spans the entire battery lifecycle, aligned with the requirements of the EU Battery Regulation and industry operational needs. The platform features a data architecture composed of 7 core clusters and over 90 data attributes, designed to comprehensively document and track battery-related information (Figure 6.6).

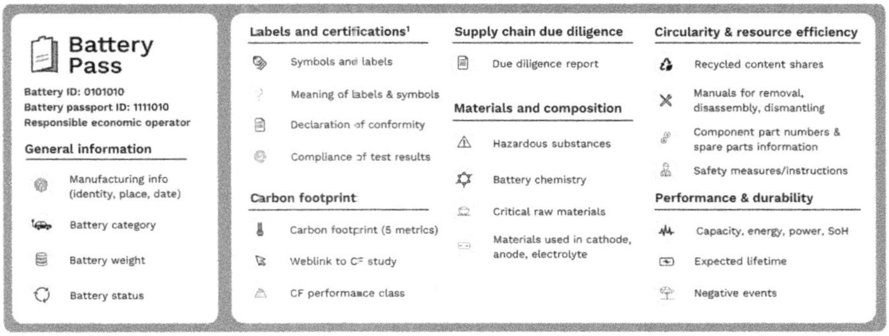

Figure 6.6 Battery Passport data content. *Source:* Battery Pass [6] / Systemiq / CC BY 4.0.

6.2.3.1 Data Architecture and Core Components

The Battery Passport system captures **static baseline information**, including:

- Unique identification (ID), manufacturing date, production location, manufacturer details, and battery type (e.g., EV batteries, industrial batteries).

  It also monitors **dynamic lifecycle data**, such as:

- Status (in use, scrapped, second life), weight, and design lifespan.

  Materials and Composition documentation includes:

- Battery chemistry systems (e.g., nickel–manganese–cobalt with the numbers as proportions of materials out of 10 [NMC811] or lithium iron phosphate [LFP] for positive/negative electrodes), electrolyte types, key raw material concentrations (lithium, cobalt, nickel), GPS coordinates of mineral sources, and supply chain compliance certifications (e.g., conflict-free mineral credentials).

  Environmental Indicators adhere to EU standards, including:

- Full lifecycle carbon footprint (compliant with the European Union Product Environmental Footprint [EU PEF] methodology), proportion of recycled materials (e.g., recycled lithium content), and resource efficiency metrics to support carbon tariff compliance and circular economy evaluations.

  Technical Parameters cover performance metrics such as:

- Rated capacity (Ah), energy density (Wh/kg), state of health (SoH), capacity degradation curves, charge–discharge cycle counts, internal resistance fluctuations, and durability (e.g., exposure to extreme temperatures or deep discharge events).

  The **Supply Chain and Compliance Module** integrates:

- Human rights and environmental risk assessments aligned with OECD guidelines, supplier audit results, and third-party certifications to ensure transparency.

Real-time Operational Data, including temperature, voltage, and current parameters, is uploaded via battery management systems (BMS), alongside records of maintenance history, thermal runaway incidents, and software updates.

End-of-life and Recycling Information provides:

- Disassembly flowcharts, hazardous substance handling protocols, recycling enterprise profiles, and material recovery rates to facilitate efficient processing of scrapped batteries.

6.2.3.2 Technical Implementation

Technologically, Battery Pass enables real-time collection of dynamic data (e.g., SoH, temperature profiles) via BMS, with monthly resolution record-keeping. Independent third-party verification by accredited bodies ensures the credibility of carbon footprint and supply chain compliance data.

Standardized API interfaces facilitate seamless integration with enterprise resource planning (ERP), product lifecycle management (PLM), and supply chain management (SCM) systems, as well as connectivity with the EU Customs Single Window and vehicle registration databases to track battery export status and scrappage markers. The platform's data aggregation capabilities generate pan-European carbon footprint distribution maps for EV batteries (e.g., 2028–2035 forecasts) and inform R&D and investment decisions through analyses of capacity degradation patterns and chemical system distributions (Figure 6.7).

6.2.3.3 Application Scenarios and Impact

The Battery Passport delivers tangible value across sectors:

- **Recycling Enterprises:** Optimizes hydrometallurgy and other recycling processes using chemical composition and SoH data, potentially reducing disassembly costs by **20–40%**.

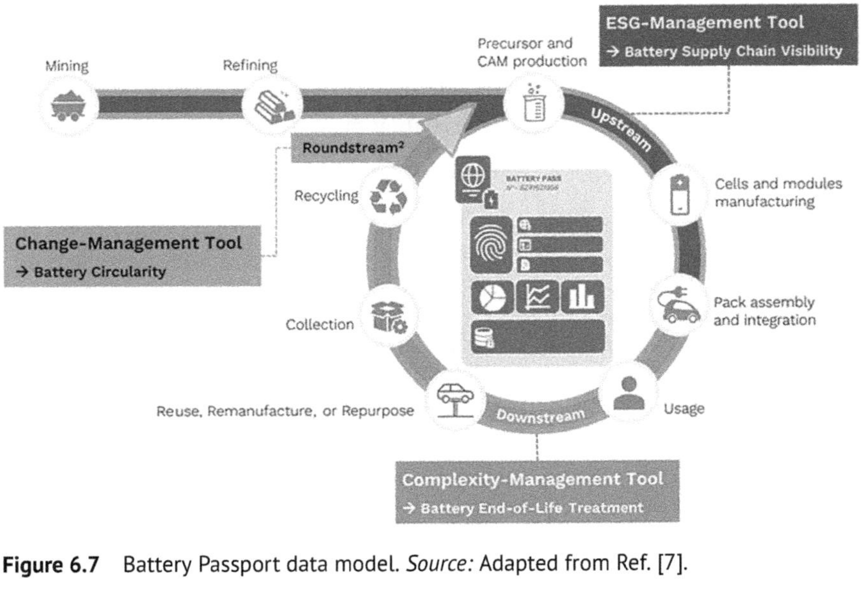

Figure 6.7 Battery Passport data model. *Source:* Adapted from Ref. [7].

- **Second-life Markets:** Establishes residual value models based on dynamic performance metrics (e.g., capacity retention, cycle counts), minimizing technical testing costs in used battery transactions.
- **ESG Compliance:** Automatically generates reports to meet regulatory disclosure requirements, while accident records and transportation risk assessments mitigate safety hazards.

6.2.4 Battery Passport Demonstrator from Battery Pass

An example of a Battery Passport from the Battery Pass project is illustrated in Figure 6.8. It showcases a Battery Passport for an EV battery, model M-41698615, with a unique Passport ID "did:web:acme.battery.pass:0228151e-949c-db67-8ef3-162431e28976" and SN 9923566100548948. Weighing 499.00 kg, it was manufactured on 5 September 2023 by Exide Batteries Auditor and verified by the economic operator NZAT.

The passport presents a summary report with a carbon footprint breakdown via a pie chart, detailing different emission sources in the battery's lifecycle. Additionally, it provides circular graphs for the recycled content share of key

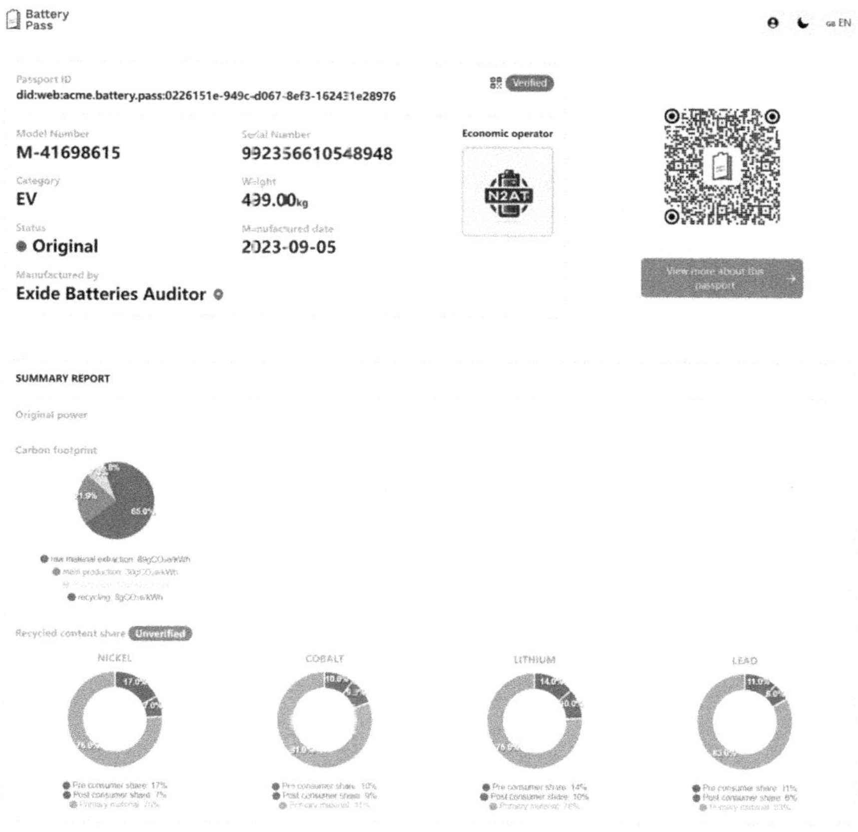

Figure 6.8 Battery Passport demonstrator from Battery Pass project. *Source:* Ref. [8] / SYSTEMIQ.

metals like nickel, cobalt, lithium, and lead, though some of which are in an "Unverified" state. This demonstrator embodies the Battery Pass project's goal of standardized battery lifecycle data management, offering a practical glimpse into how battery-related information, from manufacturing details to environmental and material composition data, is documented and can be utilized for compliance, value chain collaboration, and circular economy initiatives as per the project's objectives.

6.3 Catena-X Initiative

6.3.1 Introduction to Catena-X

Established in 2021 by Germany's Industry 4.0 platform, Catena-X endeavors to establish an open collaborative data ecosystem. The goal is to drive digital transformation and sustainable development within the industrial sector. Its membership encompasses numerous German industrial powerhouses like Deutsche Telekom, Siemens, BMW, and Volkswagen. Moreover, it includes international enterprises from diverse industries such as automotive, mechanical engineering, and chemistry.

As an open-source data ecosystem tailored for the automotive industry, Catena-X is committed to facilitating cross-enterprise data sharing via standardized interfaces. In doing so, it offers a platform for data sharing and integration for the Battery Passport. Through Catena-X's data ecosystem, every stakeholder in the battery industry chain, including raw material suppliers, battery manufacturers, automakers, and recycling companies, can attain data interoperability. Take battery manufacturers as an instance; they are able to upload data such as battery production details and raw material origins to the Catena-X platform. This enables downstream automakers and recycling enterprises to access this data with ease, thereby supporting the completion and management of Battery Passports (Figure 6.9).

6.3.2 Core Technical Capabilities

The Catena-X Battery Passport establishes a digital management closed-loop for the entire battery lifecycle by virtue of three core capabilities: data interconnectivity, compliance support, and secure sharing.

6.3.2.1 Data Interconnectivity

Based on the Eclipse Dataspace Connector (EDC), the system realizes cross-enterprise and cross-stage data integration. As the data exchange hub of the Catena-X ecosystem, EDC connects decentralized data sources in the whole process of battery production, usage, and recycling through standardized interfaces (such as IDS-RAM 3.0).

- **Operation Mechanism:** Take a battery manufacturer's manufacturing execution system (MES) as an example. It can publish raw material procurement records and carbon emission data to EDC. Automakers then initiate targeted query requests to EDC via DPP applications. EDC dynamically returns encrypted data packets

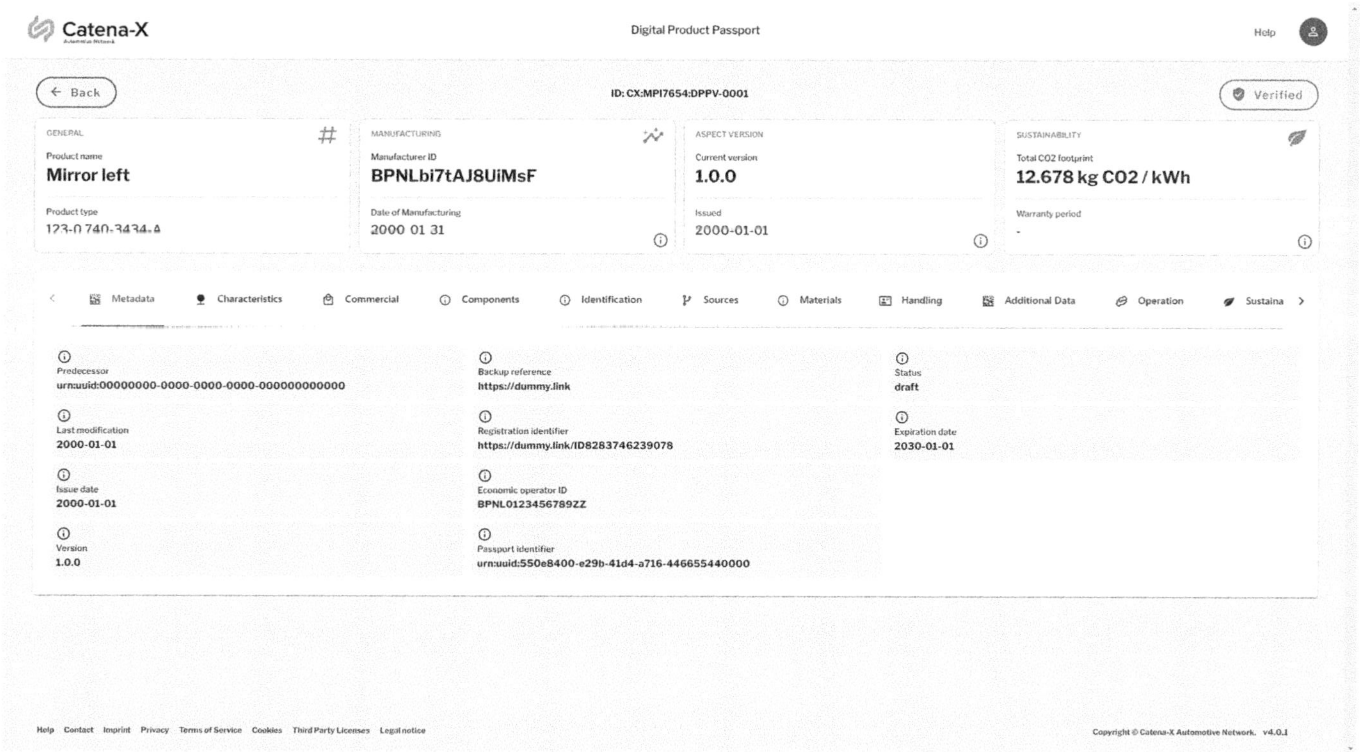

Figure 6.9 Battery Passport demonstration from Catena-X. *Source:* Catena-X [9] / Catena-X.

according to predefined permission policies, and these packets are finally parsed by the DPP backend to generate a unified Battery Passport view.
- **Data Adaptability:** This mechanism supports both structured data (such as JSON and XML) and unstructured documents, ensuring the comprehensive integration of multiformat data throughout the lifecycle.

6.3.2.2 Secure Sharing

Secure sharing is a core differentiating advantage of DPP. By leveraging blockchain and distributed ledger technology (DLT), the system ensures data security through permissioned blockchain technologies like Hyperledger Fabric or IOTA Tangle.

- **Data Integrity Assurance:** The hash values of critical data (such as the carbon emission records of production batches) are written to the blockchain, while the original data remains in the local systems of enterprises and is shared on demand only through EDC interfaces. For example, when a battery manufacturer publishes data, the system generates a hash value and stores it on the blockchain. After consumers obtain the data, they can verify its integrity by recalculating the hash value and comparing it with the on-chain record.
- **Privacy Protection Technology:** Zero-knowledge proof (ZKP) technology enables enterprises to prove compliance with regulatory authorities (such as "conflict-free cobalt sourcing") without disclosing sensitive information like supplier names, achieving the balance between compliance verification and privacy protection (Figures 6.10 and 6.11).

6.3.3 Implementation Path

Catena-X's technical architecture adheres to a strict front-end and back-end separation principle:

- **Front-end:** Leveraging Vue.js 3 and the Verify framework, it constructs intuitive user interfaces. This allows for rapid retrieval of Battery Passports through QR code scanning or manual input of part Instance IDs. Additionally, it dynamically renders standardized data models encompassing critical information such as carbon footprints and material traceability.
- **Back-end:** Built around Java and the Spring Boot framework, it integrates seamlessly with Catena-X shared services. These include energy data exchange (EDC) data connectors, a digital twin registry, and information retrieval system (IRS) component drilling services, thereby enabling cross-enterprise data interconnection.

6.3.3.1 Data Flow Architecture

The data flow within the system is structured across four core stages:

- **Discovery Stage:** Utilizes EDC to locate and establish connections with relevant data providers.
- **Digital Twin Registry:** Conducts searches and verifications of manufacturer information to ensure data accuracy and authenticity.

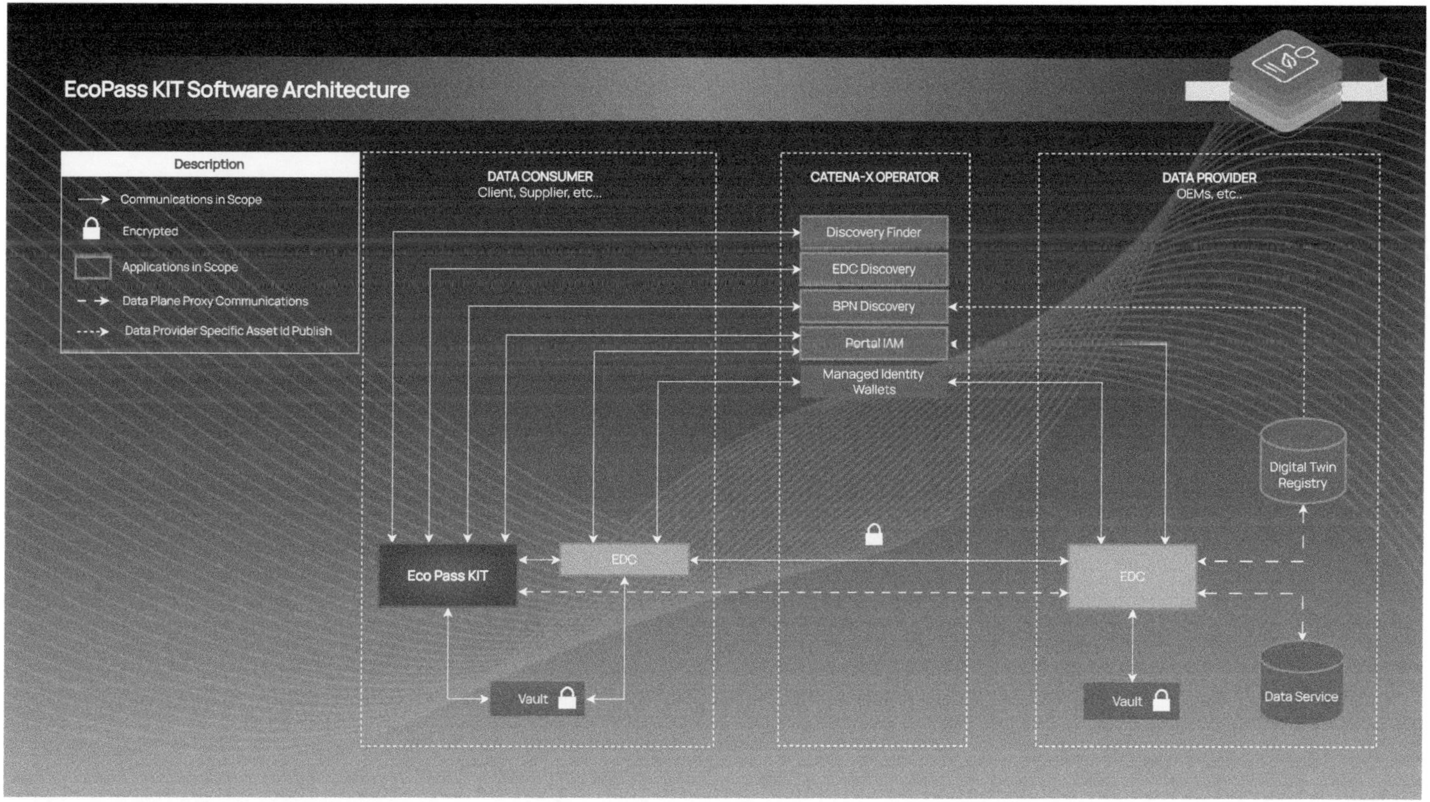

Figure 6.10 Battery Passport architecture from Catena-X. *Source:* Catena-X [9] / Eclipse Tractus-X / CC BY 4.0.

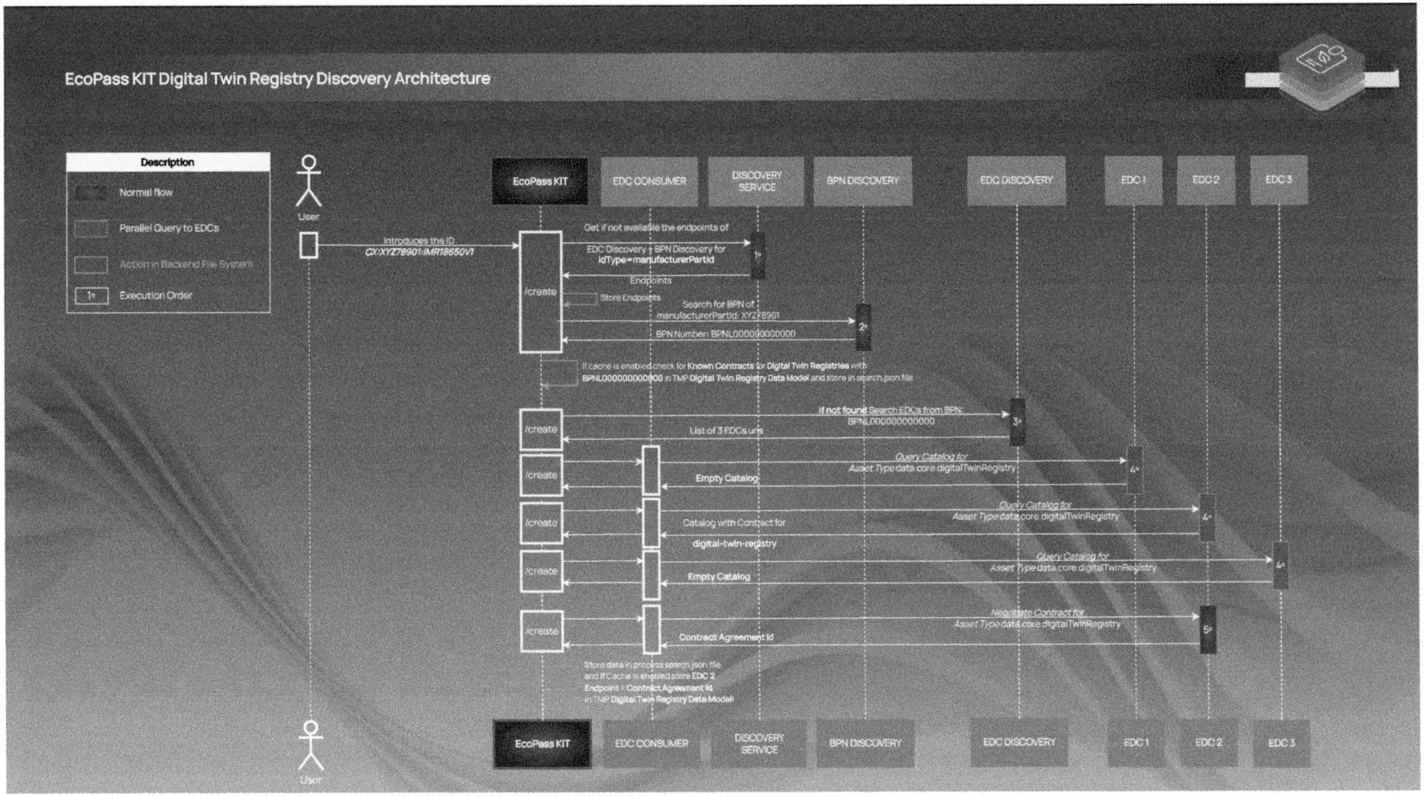

Figure 6.11 Catena-X Battery Passport front-end and back-end flow. *Source:* Catena-X [9] / Eclipse Tractus-X / CC BY 4.0.

- **IRS Integration:** Enables component-level drilling, such as mapping the relationships between battery modules and cells, to provide granular data insights.
- **Data Sharing:** Culminates in EDC negotiation and encrypted data transmission, ensuring secure and reliable data exchange between stakeholders.

6.3.3.2 Security Mechanisms

The system incorporates robust security measures to safeguard data integrity and privacy:

- **Dual-token Authentication:** Employs Keycloak to implement a dual-token authentication system, comprising user tokens and technical tokens, to enforce strict access control.
- **Encrypted Transmission:** Utilizes AES encryption for data transmission, ensuring that information remains confidential during transit.
- **ZKP:** Implements ZKP technology to protect privacy, ensuring data sovereignty and immutability while allowing for secure data verification without exposing sensitive details.

6.3.3.3 Compliance Support

To meet regulatory requirements, the system is equipped with built-in carbon emission factor libraries and algorithms. This enables it to automatically generate reports compliant with the EU Battery Regulation, supporting one-click export to regulatory platforms and streamlining the compliance process for enterprises.

6.3.3.3.1 Deployment and Operations

- **Containerized Management:** Implements containerized deployment based on Kubernetes and Helm, ensuring consistency and scalability across different environments.
- **GitOps Workflow:** Integrates with Argo CD's GitOps workflow to enable efficient and reliable application deployment and management through version-controlled configurations.
- **Monitoring System:** Deploys Prometheus/Elasticsearch, Logstash, and Kibana (ELK) monitoring systems to provide real-time insights into system performance, ensuring high availability and enabling proactive maintenance.

6.3.3.3.2 Additional Features

- **Single API Retrieval with Automatic Contract Signing:** Simplifies data acquisition processes by allowing users to retrieve data through a single API call, with automatic contract signing to streamline legal formalities.
- **SSI Wallet Verification Plugins:** Enhances data credibility by integrating self-sovereign identity (SSI) wallet verification plugins, providing the battery industry chain with end-to-end technical support for scenarios such as compliance audits and recycling optimization.

6.4 Ouranos Ecosystem and Battery Passport

6.4.1 Introduction to Ouranos Ecosystem

The Ouranos ecosystem represents a national-level data-sharing strategy spearheaded by Japan's Ministry of Economy, Trade and Industry (METI), with the overarching goal of achieving "Data Free Flow with Trust" (DFFT) through the establishment of cross-industry and cross-enterprise data collaboration networks [10]. Guided by the vision of "Society 5.0," this initiative seeks to integrate the physical and digital spaces, enhance industrial competitiveness, and align with core Japanese government policies such as the Basic Design for a New Capitalism and the Digital Society Priority Plan. Its core objective is to address global challenges, including supply chain transparency, resource circularity, and carbon neutrality, through standardized data architectures and interoperable platforms. Ouranos's "System of Systems" (SoS) architecture and its explicit focus on "Data Free Flow with Trust" (DFFT) signify a national strategy that prioritizes seamless, secure, and trusted cross-border data exchange as a fundamental pillar for economic competitiveness and addressing global challenges. DFFT is a high-level policy vision championed by Japan. Ouranos is Japan's concrete, national-level implementation of this vision. The SoS architecture implies that Ouranos is building a meta-framework designed to connect and enable interoperability among multiple, disparate data platforms and systems, rather than creating a single, monolithic data repository. This approach reflects a deep understanding that data is inherently distributed across various enterprises and jurisdictions. By focusing on foundational layers like semantics, identity, and connectors, Ouranos aims to unlock the value of data across borders, directly addressing challenges like regulatory fragmentation and lack of clear definitions (Figure 6.12) [12].

In the battery sector, Ouranos is specifically focused on developing and promoting Battery Passports to meet the compliance requirements of the EU Battery Regulation. The overarching goal is to achieve data transparency across the entire battery lifecycle – from raw material extraction to recycling and reuse. The Ouranos Battery Passport integrates three types of critical information: identification information (manufacturer, specifications, resource composition), value information (performance data, historical usage records), and environmental information (carbon footprint, supply chain due diligence, proportion of recycled materials). This comprehensive approach ensures that batteries comply with international environmental and human rights standards while simultaneously supporting circular economy and carbon neutrality goals (Figure 6.13).

6.4.2 Technical Architecture of the Ouranos Ecosystem Data Space

The Ouranos Ecosystem Data Space (ODS) employs a SoS core architecture, designed to facilitate seamless cross-industry and cross-border data sharing and collaboration through a layered technical framework and multiperspective governance model. Rooted in the principles of decentralization and interoperability, ODS

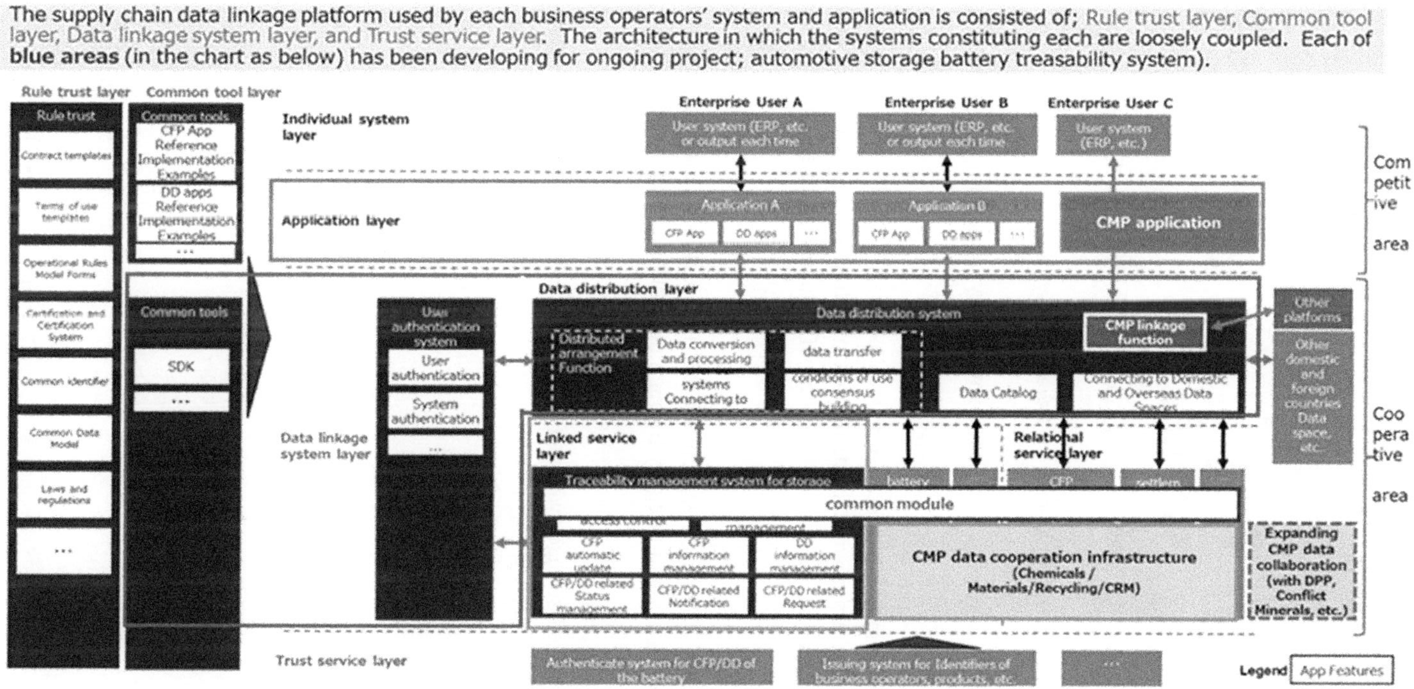

Figure 6.12 Concept of integration between Circular Management Platform (CMP) and the Ouranos Ecosystem. *Source:* Ministry of Economy, Trade and Industry [11] / Ministry of Economy, Trade and Industry.

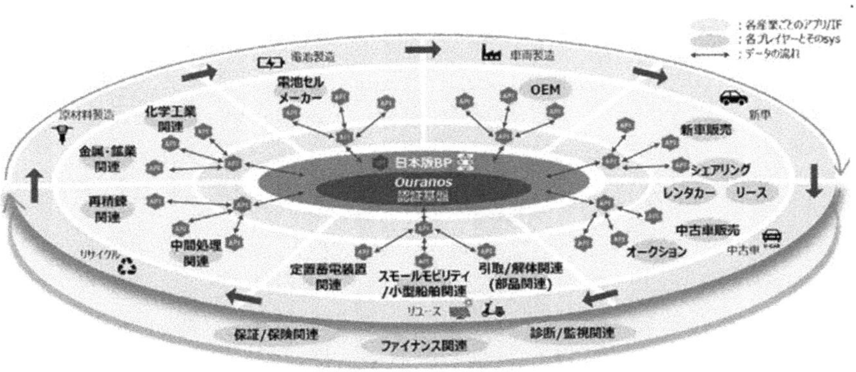

Figure 6.13 Battery Passport based on the Ouranos Ecosystem.

addresses the challenges of data silos and sovereignty management across complex ecosystems.

6.4.2.1 Technical Framework: Four-Layer Architecture

The ODS technical framework is structured into four distinct yet interoperable layers, each addressing specific dimensions of data management and exchange (Figure 6.14):

1) Data Layer (L1)
   - **Functionality:** Manages structured and unstructured data assets, ensuring **data sovereignty** and integrity through distributed storage mechanisms and metadata tagging.
   - **Key Mechanisms:**
     - Decentralized storage systems (e.g., blockchain-adjacent technologies) for secure data custody.

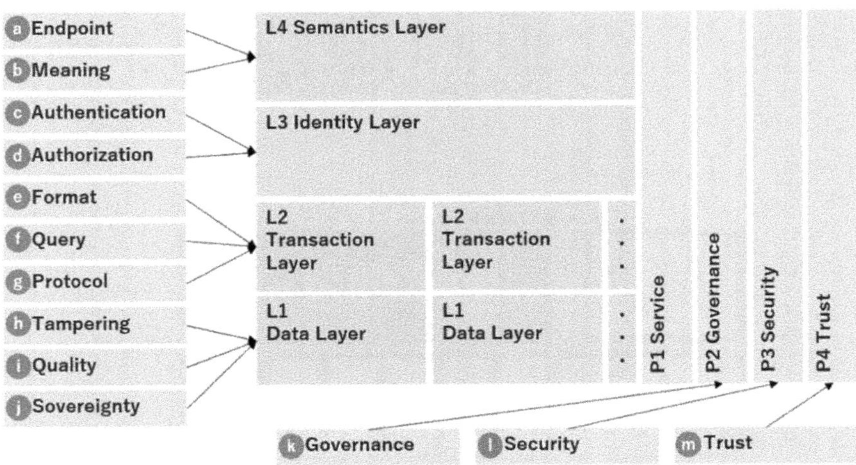

Figure 6.14 Architecture of Ouranos Ecosystem. *Source:* Ministry of Economy, Trade and Industry [13] / Information-technology Promotion Agency, Japan (IPA).

  - Metadata standardization to enable semantic enrichment and discoverability.
- **Objective:** Empower data providers with autonomous control over data access and usage conditions, aligning with the principle of self-determination in data sovereignty.

2) Transaction Layer (L2)
- **Functionality:** Facilitates reliable data transmission across diverse formats and protocols, supporting both real-time and asynchronous data exchange.
- **Key Mechanisms:**
  - Support for batch processing, streaming, and media data transfer protocols.
  - Lightweight APIs and middleware for seamless integration with existing enterprise systems.
- **Objective:** Reduce technical friction in data transfer, enabling efficient cross-organizational workflows (e.g., supply chain traceability, real-time analytics).

3) Identity Layer (L3)
- **Functionality:** Enables cross-domain identity management and access control through decentralized identifiers (DIDs) and verifiable credentials (VCs).
- **Key Mechanisms:**
  - OAuth 2.0 and OpenID Connect for federated authentication.
  - DID-based profiles and VCs for secure, tamper-proof identity verification.
- **Objective:** Establish trust across heterogeneous networks by enabling seamless identity federation (e.g., verifying supplier compliance in global supply chains).

4) Semantic Layer (L4)
- **Functionality:** Ensures interoperability between disparate systems through standardized ontologies and data mapping.
- **Key Mechanisms:**
  - Aspect Model and other semantic frameworks for unified data modeling.
  - Metadata exchange protocols to reconcile schema differences (schema-flexible approach).
- **Objective:** Enable machine-readable data interoperability, supporting AI/ML applications and cross-system analytics.

6.4.2.2 Core Components and Governance Principles

1) **Semantics Component:** Enabling Semantic Interoperability
- **Functionality:** Standardizes metadata and ontologies to bridge heterogeneous systems.
- **Use Cases:**
  - **Battery Passport Compliance:** Maps carbon footprint (CFP) data to EU Battery Regulation standards via Japan's IDEA database, ensuring transnational regulatory alignment.
  - **Geospatial Data Integration:** Utilizes 4D spatiotemporal IDs (Spatial Voxel) for real-time disaster management, enabling cross-agency coordination in shelter allocation and infrastructure monitoring.

2) **Identity Component:** Decentralized Identity Federation
   - **Functionality:** Leverages DIDs and VCs to create a trust layer for cross-data-space interactions.
   - **Use Cases:**
     - **Global Supply Chain Compliance:** Enables enterprises to verify supplier credentials via VCs (e.g., alignment with the EU's Catena-X framework), reducing audit redundancies.
     - **Federated Authentication:** Supports seamless access to multiorganizational datasets (e.g., shared mobility networks).
3) **Dataspace Connector:** Interoperable Data Exchange
   - **Functionality:** Implements International Data Space Association (IDSA) protocols for lightweight, standards-based data exchange.
   - **Key Features:**
     - Synchronous/asynchronous API supports flexible data transmission.
     - Machine-readable contract negotiation for automated access rights management (e.g., battery recyclers negotiating access to recycled material data).
     - Compatibility with cloud storage and real-time streaming platforms (Figure 6.15).

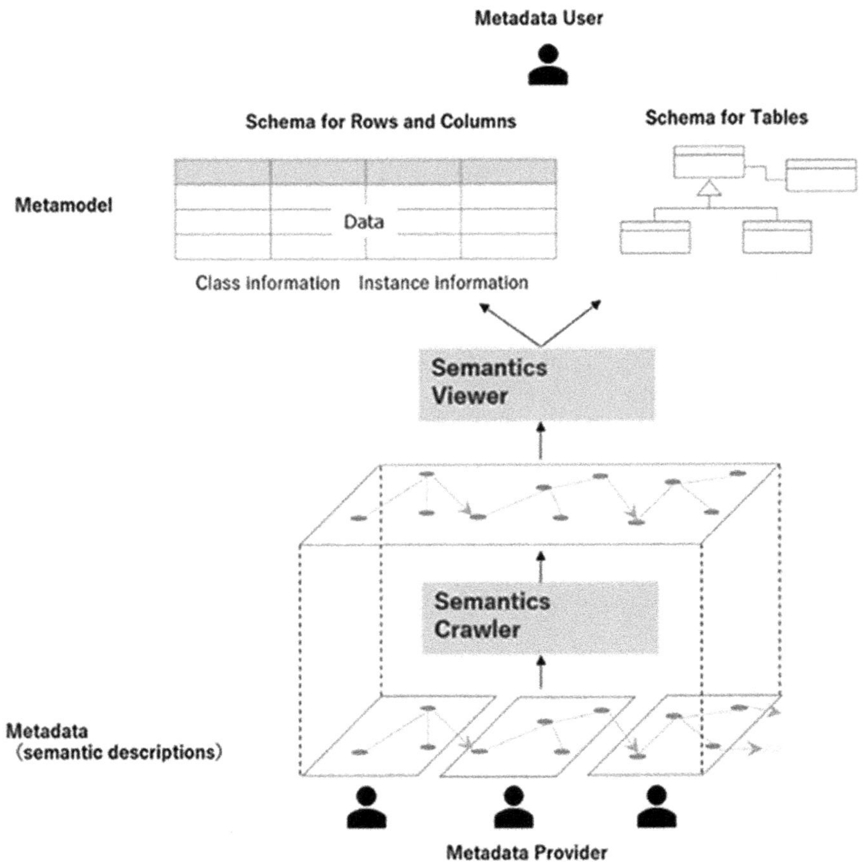

Figure 6.15 How the semantics component works. *Source:* Ministry of Economy, Trade and Industry [13] / Information-technology Promotion Agency.

4) **Governance:** Data Sovereignty and Trust Management
   - **Principles:**
     - **Data Sovereignty Priority:** Data providers retain control over data sharing parameters (e.g., granular access to battery CFP data).
     - **Privacy Preservation:** Employs differential privacy and federated learning to protect sensitive data (e.g., trade secrets in supply chain networks).
     - **International Interoperability:** Aligns with frameworks like Gaia-X and the EU's Digital Single Market, enabling cross-border data mapping (e.g., CIDIME-certified carbon data converted to EU due diligence formats).
   - **Risk Management:** Integrates AI-driven anomaly detection and third-party audit logs to monitor data quality and detect fraudulent activities (e.g., irregularities in carbon emission reporting).

The Ouranos Ecosystem Data Space (ODS) serves as a foundational infrastructure for global data collaboration, addressing the technical and governance challenges of cross-border data interoperability. By prioritizing data sovereignty, semantic standardization, and decentralized trust mechanisms, ODS enables industries to unlock the value of data while adhering to regulatory and privacy requirements. As it evolves, ODS aims to support global sustainability goals, including carbon neutrality and resource circularity, by providing a robust framework for secure, interoperable data ecosystems.

6.5 Battery ID

6.5.1 Background Introduction

In the context of the profound integration between digital transformation and the real economy, the digital economy has been experiencing robust growth. The Overall Plan for the Construction of Digital China underscores the critical role of digital trust in fostering a digital ecosystem. As a "key" for entities to access the digital realm, digital identity acts as the foundation of the digital trust system and the core enabler for data interaction. It transforms real-world identity information into a unique digital identifier, encapsulating the diverse attributes and permissions of entities within the digital domain (Figure 6.16).

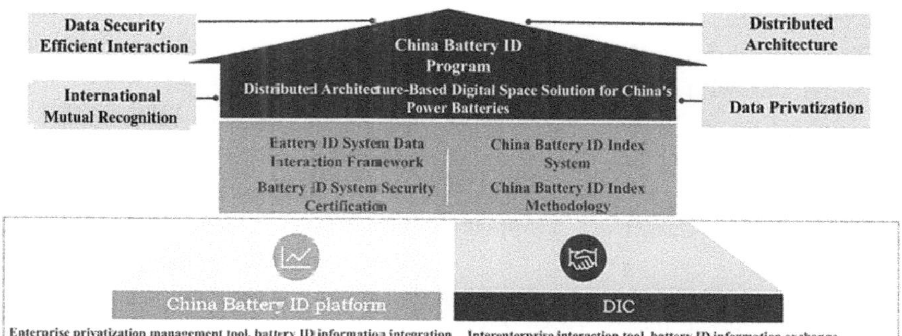

Figure 6.16 Basics of the China Battery ID project. *Source:* CATARC [14] / United Nations Economic Commission for Europe.

With the rapid advancement of the new energy vehicle and power battery industry, the digital management of the entire battery lifecycle has become an intrinsic requirement for industrial development. On 26 December 2024, China Automotive Technology & Research Center Co., Ltd. (CATARC) convened the 2024 China Battery ID Achievements Conference. At this conference, five vehicle manufacturers and five battery enterprises unveiled the first batch of China Battery IDs.

Leveraging data carriers such as QR codes, China Battery ID assigns a unique identifier to each battery. It consolidates data across the battery's entire lifecycle and establishes a specialized archive. The indicator system of China Battery ID spans six dimensions: battery information, disassembly and recycling, sustainability, manufacturers and suppliers, warning signs, and dynamic information.

Centering on indicators including basic battery information, capacity and voltage, battery life, power capability, temperature status, energy efficiency, material information, recyclability design, and carbon footprint, the first batch of enterprises to acquire digital identities has finished data collection and accounting. This has initially realized new practices and applications in the digital management of China's battery industry.

6.5.2 Technical Framework of China Battery ID

The technical framework of China's Battery ID and the implementation of China's Battery ID are rooted in a decentralized digital space solution, achieving trusted data interaction across all industrial chain links through a distributed architecture. This design ensures data sovereignty and enhances cross-enterprise collaboration efficiency. With blockchain technology as its core support, the architecture integrates enterprise-level private management tools to protect data privacy and security while enabling mutual recognition of international standards (e.g., digital signature authentication), providing a compliant foundation for the global circulation of batteries (Figure 6.17).

6.5.2.1 Core Functional Modules

China's Battery ID system comprises three core functional modules, forming a complete technical chain covering "identity generation – data services – interaction authentication":

1) Battery ID Management System
   - **Standardized Identity Generation:** Generates unique identity identifiers in compliance with national standards.
     - **New Energy Vehicle Batteries:** Utilize **24-bit codes** encompassing full lifecycle information such as production batches, installed vehicles, and recycling status.
     - **Second-life Batteries:** Adopt **19-bit codes** to clarify origins and reuse scenarios.
   - **Multimodal Label Integration:** Combines physical labels (for basic information) with electronic labels (for extended data) to achieve hierarchical information storage and rapid traceability across the entire lifecycle.

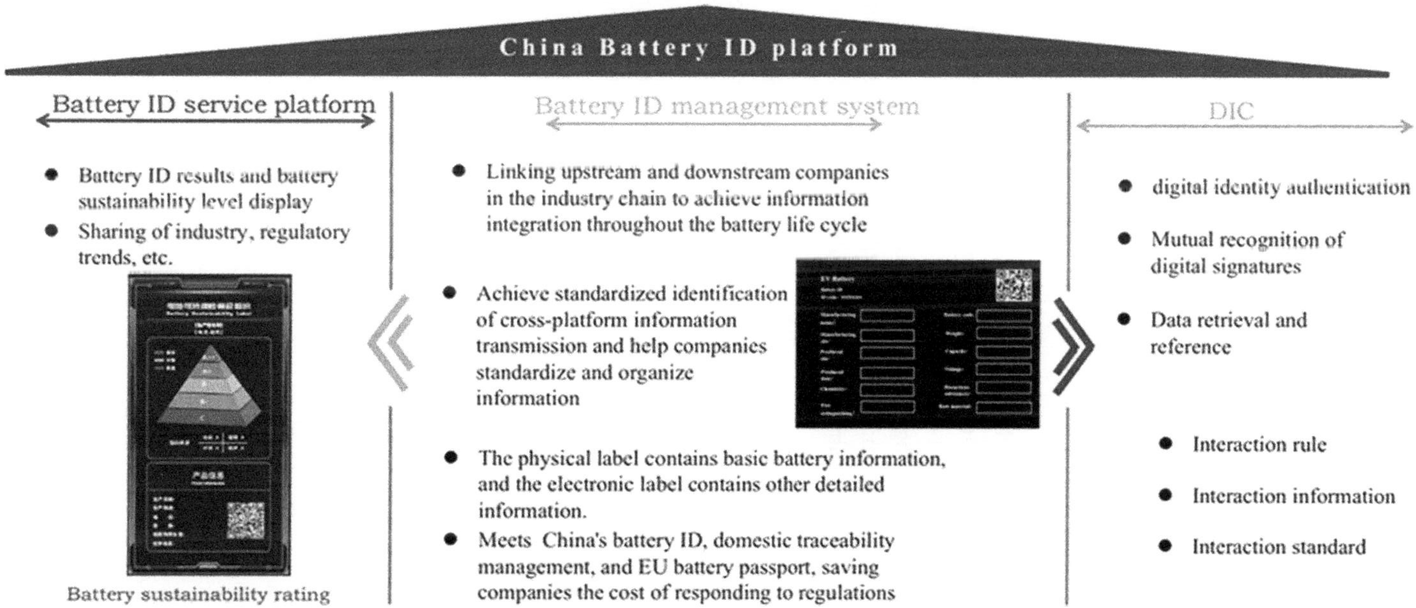

Figure 6.17 Architecture of China Battery ID. *Source:* CATARC [14] / United Nations Economic Commission for Europe.

2) Battery ID Service Platform
   - **Comprehensive Service Integration:** Provides data query, compliance certification, and industry trend analysis and integrates battery sustainability rating functions.
   - **Rating System Construction:** Generates quantitative ratings by synthesizing key indicators (e.g., carbon footprint, resource circulation efficiency), assisting enterprises in meeting domestic and international regulatory requirements and promoting the development of green supply chains.
3) Data Interaction Tools
   - **International Compatibility:** Aligns with international norms such as the EU Battery Passport through standardized protocols, enabling cross-platform information transmission and reducing multistandard adaptation costs for enterprises.
   - **Distributed Technical Architecture:** Employs a hybrid model of "centralized management + distributed storage" to ensure efficient data interaction while mitigating single-point failure risks.
   - **Security and Mutual Recognition Mechanisms:**
     - **Blockchain Application:** Leverages distributed ledgers to ensure data immutability and verifiability, complemented by multilevel permission control to safeguard enterprise data sovereignty.
     - **Digital Identity Mutual Recognition:** Relies on technologies such as DID to establish international mutual recognition, promoting interoperability between China's Battery ID and global standards.

This system provides a digital infrastructure that balances security, compliance, and efficiency for the global expansion of the battery industry through technological innovation and standard coordination, facilitating the construction of a transparent and trustworthy global battery circular economy ecosystem.

6.5.3 Example of China Battery ID

The example of China Battery ID from CATL is shown in Figure 6.18. The Indicator System of China Battery ID encompasses a comprehensive framework. It consists of six major dimensions, namely, basic information, disassembly and recycling, sustainability, manufacturer and supplier, warning sign, and dynamic information. These dimensions are further divided into 20 detailed categories, which cover aspects like basic battery information, capacity and health information, power capability, and more. To measure and assess these, there are 132 process-based indicators, 93 outcome-based indicators, and 16 methodologies, providing a holistic approach to battery identification and management.

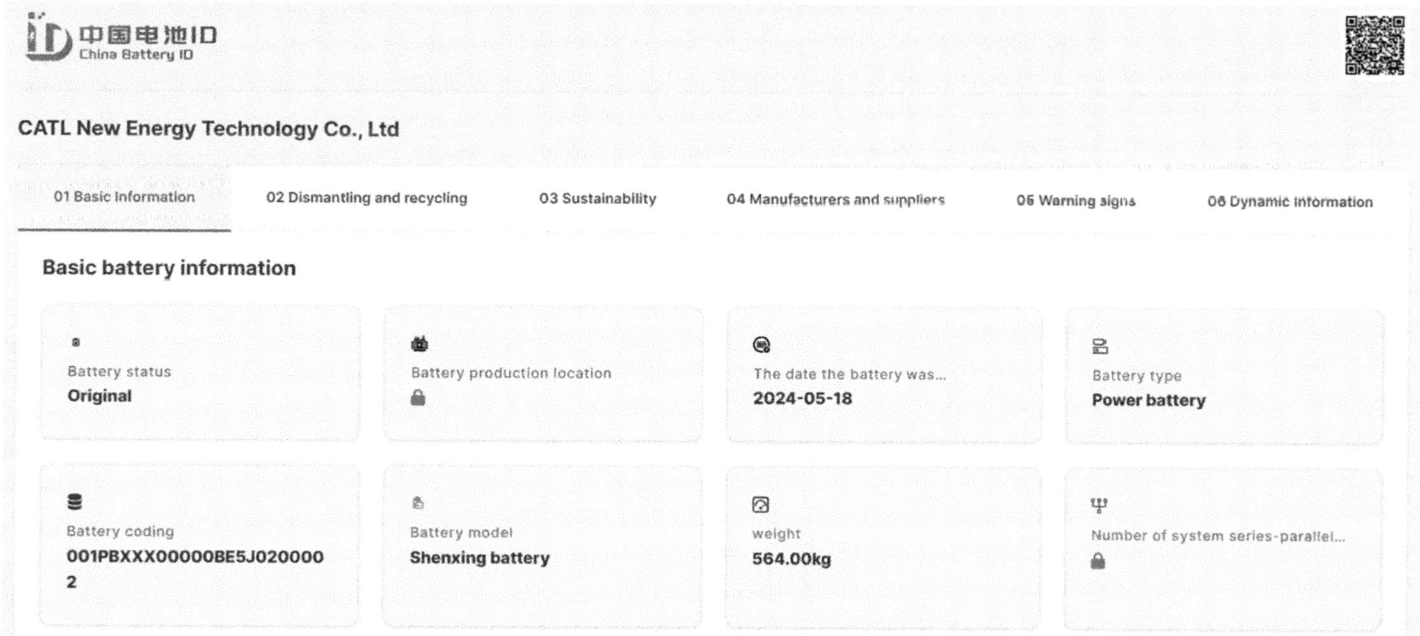

Figure 6.18 Example of a China Battery ID from CATL. *Source:* Ref. [15] / China Automotive Data Co.

6.6 Automaker and Technology Provider Initiatives

Beyond the major alliance and collaborative projects, individual automakers and technology providers are actively developing and implementing Battery Passport solutions, often in direct response to regulatory mandates and consumer demand for greater transparency. These initiatives demonstrate the practical application of Battery Passport concepts within specific product lines and supply chains.

6.6.1 Volvo's Battery Passport

Volvo Cars has emerged as a leader in implementing Battery Passports, with its flagship EX90 all-electric SUV being the first vehicle to come equipped with a Battery Passport as standard. This initiative, powered by British technology firm Circulor, is a result of a collaborative effort that began in 2019. The Volvo EX90's Battery Passport provides end-to-end primary data, offering unprecedented supply chain transparency attributed to individual vehicles [16].

Key information accessible to EX90 drivers via the Volvo Cars app and a QR code on the door frame includes:

- **Material Traceability:** Details on the origin and production journey of critical raw materials such as cobalt, nickel, graphite, and lithium. This directly addresses concerns about mining practices, particularly in regions like the Democratic Republic of Congo, where child labor and environmental impacts are significant issues.
- **Embedded Carbon Footprint:** The environmental impact of the entire battery pack.
- **Recycled Content:** The percentage of recycled material used in the battery.
- **Lifecycle Data:** Future updates are expected to include battery health data.

The implementation of Volvo's Battery Passport is driven by the EU Battery Regulation, which mandates digital Battery Passports for all EVs by February 2027. It also assists Volvo in qualifying for incentives under the U.S. Inflation Reduction Act's EV tax credit, which is tied to domestic sourcing requirements for battery minerals. While the passport adds approximately $10 to the cost per car, this investment is offset by enabling market access in the EU and unlocking U.S. incentives [17]. Both Volvo and Circulor are members of the GBA, further aligning their efforts with broader industry sustainability goals. This initiative signifies a significant step toward greater transparency and sustainability in the EV battery supply chain, setting a new standard for the automotive industry.

6.6.2 Siemens's Battery Passport

Siemens provides a comprehensive Battery Passport as a Software-as-a-Service (SaaS) solution, designed to digitally document the complete battery lifecycle. This solution enables economic operators to create, manage, and export Battery

Passports, facilitate data input from their suppliers, and ensure compliance with regulatory requirements.

The Siemens Battery Passport is built on a platform designed for broader application, extending beyond batteries to incorporate other DPPs as mandated by EU regulations such as the Eco-design for Sustainable Products Regulation (ESPR). This modular and scalable approach allows businesses to meet current and emerging product passport regulations across various industries. The Siemens Battery Passport SaaS solution is commercially available, offered in various sizes (S, M, L, XL) to match diverse production needs, and includes API access, user accounts, and a set number of DPPs to support long-term compliance and operational goals. This demonstrates a strategic move by technology providers to offer ready-made solutions that streamline the complex process of Battery Passport implementation for manufacturers and other stakeholders [18].

6.7 Conclusions

The analysis of global Battery Passport initiatives reveals a clear and accelerating trend toward comprehensive digital management of the battery value chain, driven by both stringent regulatory mandates and proactive industry collaboration. The EU Battery Regulation (2023/1542) stands out as a pioneering and globally influential framework, characterized by its holistic lifecycle approach, detailed phased implementation, and ambitious targets for environmental sustainability, ethical sourcing, and circularity [19]. Its comprehensive scope, covering carbon footprint declarations, minimum recycled content, performance standards, supply chain due diligence, and end-of-life management, positions it as a de facto global benchmark, compelling multinational corporations to adopt similar high standards to ensure market access and operational consistency worldwide.

The GBA pilots, Battery Pass, Catena-X, Ouranos, China Battery ID, and OEM initiatives like Volvo's and Siemens's each contribute distinct yet complementary capabilities to this evolving ecosystem. The GBA's iterative pilot approach demonstrates a pragmatic commitment to refining the passport framework, notably by empirically identifying and addressing critical human rights issues like child labor. Battery Pass highlights the profound technical complexity and granular data requirements for dynamic lifecycle tracking, while also showcasing tangible economic benefits such as reduced recycling costs and new business models in the circular economy. Catena-X underscores the importance of decentralized data sharing and privacy-enhancing technologies like ZKPs, addressing industry concerns about data sovereignty in competitive environments. Japan's Ouranos ecosystem, rooted in the "Data Free Flow with Trust" vision, exemplifies a national strategy to build interoperable data spaces that facilitate secure cross-border data exchange, recognizing its fundamental role in economic competitiveness and addressing global challenges. China's battery ID has attracted the participation of various local battery and automotive OEMs. Finally, the proactive adoption by OEMs like Volvo, leveraging technology providers such as Circulor and Siemens, demonstrates the practical implementation and tangible benefits of Battery Passports in meeting regulatory demands and enhancing brand reputation through transparency.

Despite significant progress, challenges persist, particularly concerning data standardization, seamless interoperability across diverse national and corporate platforms, and the delicate balance between transparency and data privacy. However, the strategic deployment of advanced technologies, such as Distributed Ledger Technologies for data authenticity, Verifiable Credentials for trusted identity management, and ZKPs for privacy-preserving compliance, has proven instrumental in overcoming these hurdles. The convergence of these initiatives, marked by planned interoperability between major data spaces like Ouranos and Catena-X, signifies an emerging global standard for trusted data exchange. This collaborative evolution suggests a future where regulatory compliance is not merely a standalone burden but is seamlessly integrated into larger, interoperable digital frameworks, fostering a more sustainable, equitable, and economically efficient global battery ecosystem. The ongoing developments underscore that Battery Passports are not just a compliance tool but a strategic enabler for the future of sustainable industrial value chains.

References

1 Global Battery Alliance (2024). GBA Battery Passport an overview. https://www.globalbattery.org/media/publications/gba-batterypassport-2024-v1-web.pdf (accessed 13 July 2025).

2 Global Battery Alliance (2023). Battery Passport 2023 pilots. https://www.globalbattery.org/battery-passport-poc-pilots (accessed 17 March 2025).

3 Global Battery Alliance (2024). Battery Passport 2024 pilots. https://www.globalbattery.org/battery-passport-mvp-pilots (accessed 24 August 2025).

4 Battery Pass (2024) Advancing Battery Passport readiness [Internet]. Berlin: Battery Pass Consortium. https://thebatterypass.eu (accessed 1 April 2025).

5 Battery Pass (2024). Battery Passport technical guidance version 1.0. https://thebatterypass.eu/assets/images/technical-guidance/pdf/2024_BatteryPassport_Technical_Guidance.pdf (accessed 29 June 2025).

6 Battery Pass (2023). Battery Passport content guidance version 1.1 (superseded by DIN-DKE-SPEC 99100). https://thebatterypass.eu/assets/images/content-guidance/pdf/2023_Battery_Passport_Content_Guidance.pdf (accessed 5 August 2025).

7 Battery Pass Consortium (2024). BatteryPass data model. GitHub. https://github.com/batterypass/BatteryPassDataModel (accessed 12 May 2025).

8 Battery Pass Consortium (2024). BatteryPass Battery Passport demonstrator [Internet]. Battery Pass. https://thebatterypass.io/did:web:acme.battery.pass:0226151e-949c-d067-8ef3-162431e28976/summary (accessed 20 July 2025).

9 Catena-X (2025). Digital product pass documentation [Internet]. Eclipse Tractus-X / Catena-X. https://eclipse-tractusx.github.io/digital-product-pass/docs/#business-statement-and-application-use-case (accessed 15 June 2025).

10 World Economic Forum (2020). Data free flow with trust: overcoming barriers to cross-border data flows [Internet]. Geneva: World Economic Forum. https://www.weforum.org/publications/data-free-flow-with-trust-overcoming-barriers-to-cross-border-data-flows (accessed 23 March 2025).

11 Ministry of Economy, Trade and Industry (2025). The Ouranos Ecosystem trust study group report [Internet]. Tokyo: METI. https://www.meti.go.jp/english/press/2025/pdf/0328_004a.pdf (accessed 9 April 2025).

12 Ministry of Economy, Trade and Industry (2023). Japan's initiatives for interoperable data infrastructures officially named "Ouranos Ecosystem" [Internet]. Tokyo: METI. https://www.meti.go.jp/english/press/2023/0429_001.html (accessed 27 August 2025).

13 Ministry of Economy, Trade and Industry (2024). Ouranos Ecosystem dataspaces reference architecture model whitepaper [Internet]. Tokyo: Information-technology Promotion Agency, Japan. https://www.ipa.go.jp/en/digital/architecture-guidelines/individual-link/d662m3000000hhu8-att/ouranos-ecosystem-dataspaces-ram-white-paper.pdf (accessed 18 July 2025).

14 China Automotive Technology and Research Center (CATARC) (2024). Chinese policy activities with respect to battery sustainability and battery ID [Internet]. Geneva: United Nations Economic Commission for Europe. https://unece.org/sites/default/files/2024-10/1_China.pdf (accessed 2 June 2025).

15 Battery ID example for CATL (2025). China: batteryid.com.cn. www.batteryid.com.cn/91350900587527783P/DIC/Pilot?id=88.186.1/91350900587527783P.PACK.1&v=2 (accessed 14 August 2025).

16 Circulor (2025). World's first Battery Passport comes on the Volvo EX90, powered by Circulor. https://circulor.com/articles/worlds-first-battery-passport (accessed 6 April 2025).

17 Politico (2024). Your next car might come with a Battery Passport [Internet]. Politico. https://www.politico.com/newsletters/the-long-game/2024/08/13/passport-please-00173795 (accessed 12 May 2025).

18 Siemens (2024). Siemens Battery Passport. Siemens Xcelerator. https://xcelerator.siemens.com/global/en/industries/battery-manufacturing/battery-passport.html (accessed 4 June 2025).

19 European Union (2023). Regulation (EU) 2023/1542 of the European Parliament and of the Council on batteries and waste batteries [Internet]. Official Journal of the European Union. https://eur-lex.europa.eu/eli/reg/2023/1542/oj/eng (accessed 1 April 2025).

Epilogue

Over the course of this book, we have explored the purpose, technology, economics, policy framework, environmental benefits, and real-world implementation of the Battery Passport. Together, these dimensions reveal a critical truth: the Battery Passport is as much a governance and economic shift as it is a technological one. It embodies a rethinking of how we track, use, and recover critical materials in an era defined by resource constraints and climate urgency.

From the outset, we saw why this system matters. Battery demand is exploding, driven by electric vehicles, energy storage, and digital technologies, but our capacity to trace where batteries come from, how they are made, and where they end up has not kept pace. This gap undermines environmental goals, fuels unethical supply chains, and generates massive waste. The Battery Passport directly addresses these issues by embedding verified data across the lifecycle of each battery, from mining and manufacturing to second-life use and recycling. The result is a system where data empowers action: manufacturers can improve sourcing, regulators can enforce sustainability, and consumers can choose responsibly.

The technological infrastructure behind the passport is equally crucial. Far from being a single platform, the system is an ecosystem combining IoT sensors, QR codes or RFID, cloud computing, and robust encryption. Each component plays a specific role in capturing, storing, securing, and transmitting battery data. But the real power lies in how these technologies interconnect to build trust, interoperability, and real-time insight. This technical backbone must be reliable, scalable, and secure, for compliance and to enable efficiency and innovation across the value chain.

Economically, the Battery Passport opens new paths. It lowers operational costs by reducing redundant testing, streamlines resource recovery, and supports regulatory compliance, offering cost savings in recycling and preprocessing. It helps companies achieve ESG goals, strengthens investor confidence, and unlocks consumer willingness to pay premiums for verified sustainable products. More importantly, it catalyzes new business models from data monetization to circular supply platforms and aligns industrial strategy with long-term planetary limits. Financial incentives and public–private partnerships will play a key role in making this vision viable, especially for SMEs.

None of this, however, exists outside a regulatory context. The Battery Passport system is being driven by the European Union's (EU) landmark 2023 Batteries Regulation, which sets a 2027 deadline for deployment in EV and industrial batteries. Other regions are moving, but unevenly. While China is investing in domestic traceability systems, North America leans on voluntary standards, and developing nations face major infrastructure and governance gaps. Without global alignment, the system risks fragmentation at the very moment it needs scale. Thus, regulatory coordination, standardization, and legal clarity remain both the backbone and bottleneck of the Battery Passport's future.

Environmentally, the case is even clearer. Battery Passports support decarbonization by making emissions transparent, enhance second-life viability by reducing testing burdens, and increase recycling efficiency by allowing for better sorting and material tracking. When fully operational, they could prevent tons of CO_2 annually and drastically reduce dependency on raw material extraction. In doing so, the passport links sustainability with competitiveness, offering a path to both climate resilience and industrial strength.

Finally, implementation shows promise but also reveals limits. Pilot programs like Germany's Battery Pass have successfully validated data templates, explored use cases from OEMs to recyclers, and tested interoperability frameworks. Yet these case studies also expose real barriers: inconsistent data quality, uneven stakeholder coordination, privacy concerns, and infrastructure mismatches. The message is clear: success depends on vision, alignment, capacity, and execution.

The Path Forward

The Battery Passport is still in its early days. Delays, resistance, and logistical hurdles are real. But its direction is set. It represents a foundational shift toward product-level accountability, data-driven circularity, and environmental integrity. Whether you are a policymaker, investor, manufacturer, or activist, the Battery Passport offers both a compliance tool and a blueprint for how to build industrial systems that respect both people and the planet.

The Battery Passport: Current Status and Key Takeaways from *The Battery Pass* Closing Event

The *Battery Pass* Closing Event brought together policymakers, standardization bodies, industry leaders, and technology providers to present findings and lessons from two years of collaborative work. Among the major outcomes was a strong consensus on the importance of aligning regulatory ambition with technical feasibility. Central to this alignment is the DIN/DKE SPEC 99100 a technical specification that provides the structural backbone for the Passport's data architecture, identity management, and lifecycle traceability.

Standardization and interoperability emerged as foundational pillars. Clear, enforceable, and internationally harmonized standards are necessary to ensure that all players, whether SMEs or multinationals, can efficiently collect, report, and share data. The Battery Pass demonstrated how data protocols, QR-based identity management, and modular technical stacks can be deployed across diverse operational contexts, paving the way for a scalable and globally compatible system.

It also emphasized the urgent need to integrate with broader Digital Product Passport frameworks, reinforcing the Passport's cross-sector value.

The project emphasized how multiple battery systems, suppliers, and jurisdictions must align their reporting mechanisms, access controls, and compliance pathways to ensure seamless data flow. Demonstrators presented at the conference ranging from cloud-based software to physical visualizations using Lego models validated the concept's practicality and adaptability. These tools showed how digital twins of batteries could be updated in real time, delivering essential information to recyclers, regulators, or OEMs at any point in the battery's life.

Business models and economic value were another major focus. Battery Passports will lower operational costs by making recycling more efficient, facilitating second-life applications, and curbing the illegal export of end-of-life batteries. They also enable precise residual value calculations for reuse, streamlining secondary markets. By 2045, experts project a 10–20% reduction in recycling costs enabled by the data accessibility the Passport provides. Companies that integrate the Passport into their ESG frameworks are expected to benefit from stronger market positioning, improved resource efficiency, and consumer trust.

Cybersecurity and data governance are critical enablers of trust in the Battery Passport system. The initiative proposed a tiered access model to ensure sensitive data such as material composition or performance metrics is only accessible to verified interest groups. Security mechanisms including cryptographic hashing, HTTPS protocols, and multifactor authentication are built into the architecture to guard against tampering and unauthorized access. Compliance with the EU's Cyber Resilience Act and GDPR is mandatory to ensure the legitimacy and integrity of the entire system.

Another central challenge discussed during the event was ensuring that small- and medium-sized enterprises (SMEs) are not left behind. Given their limited technical and financial capacity, SMEs require access to shared tools, compliance templates, and standardized data models. The Battery Pass project recommended public–private partnerships, grant schemes, and precompetitive collaboration to support SME onboarding. The event also highlighted the need for harmonized data formats across the industry to avoid duplication and inefficiencies.

A clear call to action emerged: stakeholders must begin preparations now. Regulatory compliance deadlines are fast approaching, and early engagement with data systems, standardization bodies, and supply chain partners is crucial. The final sessions emphasized the need for proactive industry leadership, investment in R&D, and coordinated implementation strategies. The European Commission reaffirmed its commitment to facilitating international alignment, particularly through forums such as the Global Battery Alliance, which is working to ensure global interoperability between national initiatives such as those in India and China.

Impact stories presented at the event underscored the initiative's feasibility. Volvo's deployment of a Passport-equipped vehicle, BASF's Pathfinder Project with Catena-X, and CEA's SURPASS 2 initiative in the electronics and textile sectors illustrated how digital product passports can be operationalized at scale. These examples revealed the potential for sector-wide transformation and provided tangible blueprints for broader rollout.

Ultimately, the Battery Passport has the potential to go beyond compliance, fostering both economic competitiveness and environmental accountability. As the battery industry becomes central to decarbonization strategies and green growth, the Passport provides the digital infrastructure needed to govern, optimize, and accelerate this transition. Through continued collaboration, standardization, and policy support, the Battery Passport can realize its role as the digital backbone of a truly circular battery economy.

Source:

The Battery Pass. Closing Event [Internet]. The Battery Pass; 2025 [cited 2025, March 12]. Available from: https://thebatterypass.eu/closing-event/

Is the Battery Passport the Right Policy at the Wrong Time? A Critical Outlook on Delays, Risks, and Industry Readiness

Following the current state of the passport as per the industry meeting at *The Battery Pass* Closing Event, we must ask if the Battery Passport is still a great idea.

While the Battery Passport initiative holds long-term promise for transparency, resource traceability, and circularity, its implementation timeline is now facing significant disruption. On 21 May 2025, the European Commission released a formal proposal to postpone the due diligence obligations tied to the Battery Regulation. Originally scheduled to take effect on 18 August 2025, these requirements central to verifying responsible sourcing are now set to be delayed by two years. This postponement, **ironically**, concerns the part of the regulation many considered the most straightforward to implement. The more technical elements of the Battery Passport especially its interoperability, real-time cloud integration, and performance tracking were always going to be more difficult. But if even due diligence cannot move forward as planned, what should we realistically expect from the rest?

The initial iteration of the Battery Passport is now unlikely to include full interconnectivity features. Instead, the system will likely launch in its most basic form, functioning as a Digital Product Passport (DPP) for batteries rather than a fully operational, real-time tracking platform. Cloud-based dynamic updates, data-sharing across international registries, and automated compliance mechanisms are all unlikely to be ready before the end of the decade. This slow and fragmented rollout is already beginning to frustrate stakeholders across the battery and automotive sectors.

Amid these delays, a growing number of stakeholders appear increasingly disoriented by shifting timelines and by the sheer complexity of the regulatory and technological landscape. SMEs lack clarity on compliance steps. Industry bodies are issuing conflicting interpretations. Tech providers are building tools without clear integration standards. Even national authorities disagree on how readiness should be measured. The result is confusion, misalignment, and paralysis – hardly the foundation needed for successful system-wide deployment.

These delays raise uncomfortable but necessary questions about timing and strategic priorities. Europe's battery and automotive industries are currently under extreme pressure. According to McKinsey (2024), the European automotive sector has lost more than 13 percentage points in market share since 2017, with profitability plummeting from 7.4% to 5.1% in just six years. EV demand is slowing, capital is scarce, and overregulation is emerging as a key threat to competitiveness. Overlaying an ambitious, highly technical compliance regime like the Battery Passport on this fragile foundation risks exacerbating existing problems rather than solving them.

At a time when European OEMs and suppliers are struggling to stabilize their operations, it is fair to ask whether now is truly the moment to introduce sweeping new digital and regulatory obligations. While the goal of the Battery Passport is admirable, its implementation could increase the cost of EVs in the short run, especially for European manufacturers already facing steep price competition from Chinese firms. In practice, the economic burden of compliance might be shifted indirectly to non-EU importers, but such measures are unlikely to materially rebalance the cost gap in Europe's favor. On the contrary, they could fuel further fragmentation, geopolitical tension, and trade retaliation.

This is not an argument against the Battery Passport as a concept. The initiative has tremendous potential to support circularity, carbon tracking, and ethical sourcing, all of which are crucial for energy security and climate goals. As part of a broader regulatory framework that includes recycled content targets, carbon footprint declarations, and export controls, the Battery Passport could help retain critical raw materials in Europe and promote cleaner production systems.

However, the deeper issue is not the policy itself but the context in which it is being enforced. The European automotive industry is undergoing structural upheaval from supply chain disruptions to skills shortages to delayed battery gigafactory investments. The industry cannot afford to lose more ground, and yet the current regulatory momentum threatens to do just that. As McKinsey underscores, more than €370 billion in gross value added is at risk by 2030 unless policymakers and industry leaders find a way to regain competitiveness and reverse the trend of decline.

By overloading the system with complex digital obligations while offering little real support or flexibility, the EU may be unintentionally deepening the challenges facing the very industries it seeks to transform. SMEs are already overwhelmed. Large players are under pressure to cut costs and offshore operations. And consumers are not yet ready to pay a premium for traceability when EV affordability remains a major concern. Instead of helping the industry recover, the current trajectory could push Europe's automotive and battery sectors even further toward irrelevance.

The Battery Passport should not be abandoned. However, it must be adapted to reality. That means acknowledging delays, simplifying early-stage requirements, offering tangible financial support for compliance, and aligning industrial policy with actual industry needs. Otherwise, we risk creating a technically brilliant framework that never gets adopted because the ecosystem it was designed to serve no longer exists.

The Battery Passport Implementation Challenge

While the policy's intent to ensure transparency, improve circularity, and reduce environmental and human rights harms is widely supported, the gap between vision and implementation remains substantial.

Technically, the passport demands the aggregation, verification, and continuous updating of vast and complex data sets across a global and fragmented supply chain. Standardization is still under development, real-time data updates remain difficult, and many companies, especially SMEs, lack the infrastructure to integrate blockchain, encryption, or interoperable APIs at scale. Without harmonized standards and middleware solutions, achieving consistent, high-quality data input from mining to recycling may be impossible. These technical burdens are compounded by a fundamental issue of data integrity: unless mechanisms such as third-party audits or automated validation are institutionalized, there is a high risk that inaccurate or manipulated data could degrade the passport's credibility, what some stakeholders call a "garbage-in, garbage-out" problem.

Regulatory enforcement adds further complexity. The passport requires dynamic updates throughout a battery's lifecycle, but there is little clarity about who is responsible for these updates during second-life use, repurposing, or recycling. Without clearly assigned roles and enforcement mechanisms, key information may go missing. Oversight authorities must also develop the capacity to audit digital passports, detect misreporting, and manage disputes over access rights, all while securing a centralized EU registry from data breaches. Ambiguity around terms like "legitimate interest" may result in data-sharing loopholes, while the heavy reliance on company self-reporting could open the door to greenwashing and undercut consumer trust unless robust verification protocols are enforced.

Industry stakeholders raise specific and recurring concerns. Battery manufacturers and OEMs are apprehensive about exposing proprietary data, facing elevated compliance costs, and navigating unclear interoperability requirements across global markets. Many see the timelines, especially the 2027 deadline, as unrealistic given the lack of finalized technical standards, unresolved access frameworks, and pending infrastructure development. Even if large firms can adapt, SMEs who often lack IT capacity and budget may be excluded from the EU market altogether unless financial aid, simplified templates, and shared compliance platforms are urgently developed. Failure to accommodate SMEs risks consolidating the battery industry around a few dominant actors, stifling innovation and undermining inclusivity.

End-of-life operators like recyclers and second-life firms broadly welcome the passport in principle but point to operational gaps in data completeness, clarity on their input roles, and user-friendly access to the system. Without reliable upstream data and streamlined interfaces, the promised benefits for recycling efficiency and safety may not materialize. Meanwhile, data privacy experts warn that even if personal data is not the main focus of the passport, risks remain. Any tracking of battery usage patterns especially in consumer-facing devices must comply with GDPR and ensure there is no scope for hidden surveillance or unauthorized marketing. Additionally, cybersecurity must be prioritized: the Battery Passport will contain sensitive supply chain and proprietary data that could become a target for hacking or industrial espionage. The entire system's integrity depends on encryption, access control, and secure architecture.

At the international level, interoperability poses a major structural challenge. The battery supply chain is global, and unless the EU actively coordinates with initiatives like the Global Battery Alliance (GBA) and aligns its standards with efforts in the United States, China, and elsewhere, manufacturers could face multiple, incompatible compliance regimes. Without such alignment, the passport could become a trade barrier rather than a catalyst for global transparency. There is a pressing need for equivalence agreements, open APIs, and standardization bodies to ensure data portability and reduce compliance duplication. While the EU has signaled interest in international harmonization, practical implementation remains a work in progress.

Above all, the passport's ultimate justification, its environmental and ethical impact, remains aspirational. Disclosing the carbon footprint, sourcing origin, or recycled content of a battery does not guarantee that these values will improve. There is a real risk of the system being exploited for greenwashing if companies are allowed to self-certify high-performance claims without independent scrutiny. Even accurate disclosures do not automatically lead to better practices unless they are tied to consequences such as procurement policies, regulatory limits, or market-based penalties. Simply stating that a battery includes cobalt from a high-risk area, or has a high carbon footprint, does not prevent its sale under current rules. If the passport remains a disclosure mechanism without regulatory teeth, it may do little more than shift reputational risk onto consumers and watchdogs without correcting the underlying problems.

Moreover, the battery passport alone cannot create circularity or ethical sourcing by itself; it enables it, but only if downstream actors like recyclers, consumers, investors, and regulators use the data effectively. If markets for second-life batteries remain weak or recycling technologies uneconomical, the passport may sit atop a broken system it cannot fix. Its value, therefore, depends on how actively it is embedded into a wider ecosystem of enforcement, innovation, and market reform. The EU must be prepared to monitor compliance and to translate insights from the passport into future performance benchmarks and regulatory restrictions; otherwise, the project risks being a well-intentioned data repository with limited real-world consequence.

The Battery Passport is both a technical system and a policy experiment, one that attempts to fuse digital infrastructure with environmental governance at an unprecedented scale. Its ambition is undeniable, but so too are its vulnerabilities. If the project is to succeed, the EU and its partners must act swiftly to finalize standards, provide support for small actors, ensure global interoperability, and design rigorous oversight mechanisms. More than anything, they must ensure that data transparency leads to accountability and that disclosure leads to change. Whether the Battery Passport becomes a transformative tool or a bureaucratic artifact will depend on the choices made in the critical years ahead.

Index

d

e